# Science Under the Microscope

## A Philosophical Investigation

## Dr. Gerard Verschuuren

En Route Books and Media, LLC
Saint Louis, MO

# ⊛ENROUTE
### Make the time

En Route Books and Media, LLC
5705 Rhodes Avenue
St. Louis, MO 63109

Cover credit: Gerard Verschuuren

Copyright © 2023 Gerard Verschuuren

ISBN-13: 979-8-88870-060-0
Library of Congress Control Number: 2023941379

No part of this book may be reproduced, stored in a retrieval system, or transmitted in any form, or by any means, electronic, mechanical, photocopying, or otherwise, without the prior written permission of the author.

# Table of Contents

PREFACE ................................................................... v
CH. 1 THE SECRETS OF MATTER ............................... 1
    The realism debate ................................................ 2
    Small causes with big consequences ..................... 4
    A fruit of evolution? .............................................. 6
CH. 2 THE WORLD AS CLOCKWORK ........................ 9
    A little tour ............................................................ 9
    Searching to find .................................................. 12
    No hidden causes ................................................. 13
    Body and soul ...................................................... 14
    God the clockmaker? ........................................... 17
    A harmonious creation ........................................ 19
CH. 3 DOES MOTHER EARTH HAVE A HEARTBEAT? ... 23
    Masters of seamanship ........................................ 24
    A natural norm? .................................................. 25
    Risk spreading ..................................................... 26
    A super organism ................................................ 27
    Where is the proof? ............................................. 29
    At the cradle of Gaia ........................................... 31
CH. 4 THE SECRET OF LIFE ....................................... 33
    Extremists ............................................................ 33
    A timeless conflict ............................................... 35
    No secret powers ................................................. 37
    The secret of life .................................................. 38
    Two-way traffic ................................................... 40

## CH. 5 THE ORIGIN OF LIFE .......... 43
### The basic unit of life .......... 44
### The fatal blow .......... 45
### A decisive experiment? .......... 49
### A primal cell? .......... 52

## CH. 6 THE EVOLUTION OF LIFE .......... 57
### The great coincidence .......... 58
### Preferred coincidences .......... 60
### Evolution on a large scale .......... 62
### The script of evolution .......... 65
### One last pitfall .......... 66

## CH. 7 IS MAN STRIPPED OF HIS CLOTHES? .......... 69
### Knocked off Man's throne? .......... 70
### Science undresses .......... 70
### Mapping the world .......... 72
### Man is nothing but... .......... 73
### Which destination? .......... 74

## CH. 8 APE OR ADAM? .......... 77
### Creationism .......... 79
### Evolutionism .......... 81
### Formidable opponents .......... 83
### What about Adam and Eve? .......... 86

## CH. 9 IS EVERYTHING IN THE GENES? .......... 89
### An a priori way .......... 90
### Yet it remains tempting… .......... 94
### Some pitfalls .......... 96
### One more case .......... 99

Table of Contents

## CH. 10 FROM SEX TO GENDER ................................................. 107
Sex ................................................................................................ 107
Gender ........................................................................................ 109
The gender myth ....................................................................... 112
The mutilation ........................................................................... 116

## CH. 11 CAN ANIMALS TALK? ..................................................... 119
Using signals ............................................................................. 121
Using labels ............................................................................... 123
Using words ............................................................................... 125
Using sentences ........................................................................ 127
Anthropomorphism ................................................................... 130

## CH. 12 CAN MACHINES THINK? ................................................ 133
The machine in Man ................................................................. 133
The brain as a machine ............................................................ 136
The brain works with concepts ................................................ 142
Artificial intelligence ................................................................ 145
Man in the machine .................................................................. 149

## CH. 13 A MATTER OF FACT ........................................................ 153
The hard facts? ......................................................................... 153
The naked facts? ....................................................................... 158
Mere observation? ..................................................................... 160
Can facts be separated from values? ...................................... 163

## CH. 14 THE GOD OF FACTS AND VALUES ............................... 173
What concepts are not .............................................................. 173
The Divine Intellect .................................................................. 176
Objective knowledge ................................................................ 178

## INDEX ............................................................................................. 181

# Preface

Every culture has its own sacred cows. In our culture, that sacred cow is science. What scientists claim—in the name of science and with the authority of science—is oracular. We are left alone in awe.

In this book, science has its say. But not with the pretension that science has the last word. We want to look beyond the scientific nose. And for that you need a philosophical analysis. That amounts to putting science under "the microscope of philosophy." I chose the image of a microscope to make visible what is invisible to the naked eye.

In each chapter, the book studies a scientific issue with an analysis that cuts to the bone. The aim is to make what scientists do and don't do a little more transparent. In short, the book offers a vision that is sometimes new, sometimes different, sometimes daring, sometimes unexpected, but—we hope—never boring.

A small selection of the book's topics may be enough to make you curious. How about DNA as the secret of life? How human is science? Was Adam forced to succumb to Mother Ape? Can animals think? And how does the machine of the human body think?

These are questions that science seems to have answered already. Or maybe not? Sometimes you must turn science inside out to find the real answers. That calls for a philosophical examination.

One more remark before you start going through this book. You can read each chapter on its own, independent of the others.

This way you can choose which chapter you want to start with because it interests you the most.

# Chapter 1

## The Secrets of Matter

What do we know about the world around us? In a way a lot, namely much more than was known hundreds or even decades ago. In another respect, what we know more completely nowadays is somewhat disappointing. In quantum mechanics, for instance, we encounter a number of situations for which there is not yet a satisfactory explanation.

That seems like a contradiction, but this contradiction is biologically quite understandable. Our knowledge may not be perfect, but in practice the knowledge we do have simply "works'—in other words: we can use it to intervene in a targeted manner. That is not surprising, because in the course of evolution the senses have been "tested" for their ability to receive information properly, and the brain has been "trained" to process this information in such a way that a response is adequate and often quite effective. A wrong reaction can be fatal, and that is precisely the target point for the mechanism of natural selection.

In short, in evolution, behavior is selected according to the degree to which it is successful. Well, successful behavior is partly based on what an organism "knows" about its environment. But whether this knowledge also corresponds to the details of the environment itself is irrelevant as long as the resulting behavior is adequate and successful. An organism's image of its environment is good enough if it does not lead to behavior with a negative selection value. That is what nature asks.

Science demands more—at least most scientists demand more than a theory that "works." In the pure sciences, also called the fundamental or academic sciences, scientists desire more than a theory that enables successful prediction and control. In that kind of sciences, people are looking not only for a successful but also truthful picture of the world around us. For example: does light consist of waves, or rather of particles? It doesn't matter if you want to see light. But to understand light it does.

### The realism debate

This brings us to the so-called realism debate. At the heart of this philosophical debate is whether the world is really as we imagine it to be? Or is such an ideal representation, given our biological past, a bit too ambitious? It is a philosophical issue that most scientists have a clear stand on. They assume that there is indeed a world outside of us, independent of us and open to scientific discovery. That's why we can distinguish facts from fictions, realities from illusions, and opinions from truths. Otherwise, why bother with science? If there is no real world, then science would be a mere hallucination.

Albert Einstein always protested vehemently against such an outcome: "The belief in an external world independent of the perceiving subject is the basis of all natural science." Fortunately, science has provided us with methods to increase our options. Nowadays almost no one dares to doubt the existence of atoms, thanks to many new observation techniques. Nevertheless, there have been some dissenting voices in the field of quantum physics. The work of quantum physicist Niels Bohr, for instance, seems to tell us that reality does not exist when we are not observing it. During one of their conversations, Bohr remembers that "Einstein suddenly stopped, turned to me and asked whether I really believed that the moon exists only when I look at it." So the question is: Are the theories of quantum mechanics a truthful reflection of reality?

To cut a long story short, it seems that, independently of perception, a physical reality does exist—as realism claims—but this reality appears to change as soon as we delve deeper. Let's take the electron as an example. The spin of an electron can be positive or negative. Only after a measurement do we know, so the claim goes, whether the spin is in fact positive or negative. Until the moment of the measurement, we are therefore faced with an uncertainty, preventing us from getting to reality itself.

This uncertainty can be explained in at least two ways. The interpretation à la Einstein is that an electron does indeed have a certain intrinsic spin, but that this also depends on a number of variables that are still hidden from our view. The rule is: Every event has a cause! We may know more later, but at this moment we must guess. So, if Einstein is right, there would be strict causality and unshakable determinism, even in the quantum range; uncertainty would only be a matter of incomplete knowledge. In principle, the underlying reality is "given" and accessible to us, says this interpretation. If only we can be patient…

Recently—partly on the basis of ingenious experiments—most physicists have opted for a different interpretation, going back to Niels Bohr. This says that an electron only has a spin once we measure it. An electron that is not measured is in a "superposition" of two states—that is, an unclear mixture of two possibilities, which can be described mathematically in a wave function. We only get certainty after measuring the spin. According to this latter interpretation, then, there is no independent reality with strict determinism and an unshakable causality, irrespective of whether we perceive that reality or not. What we perceive changes through this very observation. A world independent of our perception may be there, but not accessible to us all by itself.

### Small causes with big consequences

This latter interpretation seems inevitable, unless... Unless the observer is also fully determined so that it is already predetermined when he or she will measure and what the result will be. If all causes do indeed have fixed effects, then our future should be completely in the iron grip of the past. In that case, Einstein would be right after all. But all those who want to accept this consequence—insofar as there is still something to be "wanted"—have thus withdrawn themselves further from the realism debate, because arguments and experiments are then no longer relevant.

Yet there is something curious about our talk of causes and effects. We humans, including the scientists among us, say it so easily: "Every event has a cause." But what kind of statement is that?

Are you saying something like "Every bachelor is unmarried"? No, because that last statement says nothing about bachelors, but at most something about our language. Anyone who knows their language well knows that we should not call someone who is married a bachelor. You can find something like that in a dictionary; you don't need much life experience for that. But with "Every event has a cause" it is different. In such a case, a dictionary won't help us much. We need a little more life experience.

What then does the statement "Every event has a cause" mean? It is something like "Every bachelor is a party boy." You only know whether bachelors are party boys when you are a bit older and keep your eyes open. Maybe it's true, maybe it isn't. Isn't that also the case with causes and their effects? Shouldn't you have seen a lot of cases to know they all have causes? Yes, at first glance yes, but on closer inspection no. After all, what kind of research would you need to find out whether every event has a cause? Suppose you encounter an event that has no cause. What then? Are you now telling everyone that not all events have a cause? They'll laugh in your face, or else they'll tell you to keep looking. Imagine that the doctor tells you that your illness has no cause. You will tell him: keep searching, doctor, because every event has a cause! He may then come up with the wrong cause, but he cannot come up with saying "there is no cause."

In short, there is something curious about the statement "Every event has a cause." The philosopher Immanuel Kant pointed out this problem long ago: a statement like this relates to our experience yet does not arise from our experience. This seems to be a typical philosophical problem, but according to certain scientists it can be solved biologically. For this, they developed the evolutionary theory of knowledge.

## A fruit of evolution?

This theory tries to explain the development of our knowledge with the principle of natural selection rooted in the theory of evolution. According to this biological theory, Kant's philosophical view is correct, but only to a certain extent. As an individual, all people have a certain type of statements stored in their knowledge arsenal. One is "Every event has a cause." Although such a statement relates to our experience, we hardly had to learn for it. We "knew" it before we experienced it. It is, as it were, part of our knowledge equipment. As a human being, you no longer need to learn that all events have causes. That's how people "look" at the world. They are hard-wired that way, so to speak.

So much for the story at the level of the individual. But at the species level, the situation is different. Kant would be wrong on that level, because even "ready-made" knowledge has indeed been built up on the basis of "experience" in the course of evolution. So, it's a matter of trial and error. Natural selection has slaughtered the "bad" patterns of knowledge. And the "good" ones remained. One of those good thinking patterns was the idea of "Every event has a cause." It turned out to work wonderfully well in nature.

In that sense, humanity as a whole would have gone through a "learning" process, and the result of this has since been stored in each individual's genetic material—that is, (prior) knowledge in the form of innate structures and ideas. Such structures are the result of natural selection and adaptation. We have inherited them from our ancestors.

This is roughly the story of the epistemological theory of knowledge. This theory would explain why our (prior) knowledge "fits" so well with the world around us. This "inside knowledge" has been tested over a long period of time. But it also comes with limitations. Evolution has tested our prior knowledge in our daily living environment, which is the macro domain of plants and animals. But the knowledge principle of causes and effects has never been tested in the micro-world of quantum mechanics, for instance, which we have only known since science opened up the range of elementary particles to our experience. Perhaps the cause-and-effect principle, so useful in the macro world of living beings, is less applicable in the micro world of elementary particles.

And that brings us back to the realism debate: is the world really as we imagine it? Especially Charles Darwin, the father of the theory of evolution, had already asked himself how reliable human thinking is, if this thinking is the product of natural selection. Perhaps we should be a little more modest. The idea that only biology tells us how we know things may be quite arrogant. Even science taken in a more comprehensive sense may not have all the answers for us. There is more to it. First of all, science never ends. There is always more to know, because the magnitude of the unknown is… well, unknown. Science is, by its very nature, a work in progress. Second, science does not offer the only way of knowing—there are other ways. Science does not and cannot have answers to *all* our questions. As it has been famously said, "gravitation is not responsible for people falling in love." That's why we need to counteract scientific triumphalism and its monopoly claims.

# Chapter 2

## The World as Clockwork

Humans are place-bound creatures. They always look at things and matters from a certain position, angle, or perspective. Through training and upbringing, people may be able to walk around something in their minds and thus highlight several sides of it, one after the other. But that never works for all sides at the same time. Bilocation is not given to us, humans. Even when we walk around an object, we are always tied to a certain position and therefore a certain angle or perspective. We do not have an overview of the whole.

We can also formulate this in a more general and philosophical way. Our world lends itself to many interpretations that can coexist. Each interpretation presupposes a certain context, a certain model, a certain angle, or a certain perspective. It is also sometimes said—in line with the philosopher Kant—that as observers we always wear a certain pair of spectacles. True, but then I would like to add that those are not permanent; we can also take up other glasses. Different glasses provide a different view of things.

*A little tour*

Let's apply this fact to a very concrete and everyday object: an old-fashioned alarm clock. In my mind, I would now like to take you on a more abstract tour around this analog alarm clock. In doing so, I choose some general perspectives that are typical of the way in which people approach the world around them. From those general angles, let's look at the jump of the second hand of the clock. What kind of glasses can we put on to look at that turning pointer or needle?

The first angle is that of causality. Anyone who puts on the spectacles of causality is looking for causes. The question is, for example, what the cause is of the second hand's movement. If you look at the alarm clock from this point of view, you will discover an entire mechanism of cogs and springs behind the moving hand. Every cause is a link in a chain of causes and effects.

Fortunately, it is also possible to take off the glasses of causality and choose a different angle, as there are also other "sides" to something like an old-fashioned alarm clock. One such possible angle is that of functionality. In this case we are not looking for a cause, but for an effect that is effective or successful. A functional alarm clock is constructed in such a way that its hands rotate. Such an alarm clock is a successful product in our culture. Likewise, a heart that pumps blood around in the body is a successful product in nature. Anyone who chooses the angle of functionality is only interested in the successful effects of a mechanism, rather than in its causes. There are also causes involved, but as long as we have the glasses of functionality on, we are only interested in the successful results; they must meet certain requirements. An alarm clock whose hands do not turn is a bad product.

## Chapter 2: The World as Clockwork

But we are not yet done with an alarm clock. The glasses of functionality can also be exchanged for the spectacles of intentionality. With those glasses on, we start looking for something other than causes and functions, namely intentions or reasons. Behind a heart pumping blood there may not be a well thought-out design, but behind an alarm clock with the hands rotating, there is an intention of someone who wanted to design a device to tell the time. The intention of the clockmakers who designed this mechanism is not a hidden link in the alarm clock itself, but it is the explanation behind the existence of this entire mechanism. The clockmaker's intention is, as it were, expressed in the mechanism of the alarm clock. Causes, functions, and intentions are not mutually exclusive, but each sheds light on its own side of the matter.

Yet another side of the matter may come from the point of view of religiosity. Anyone who puts on the spectacles of religiosity goes in search of the greatest whole within which we live: the meaning of all that exists. It is the classic question from the old catechism: what are we on earth for? Or more philosophically formulated: why is there something rather than nothing? What is the basis of all that exists? Let's go back to our alarm clock again. Looking at a clock from a religious point of view has not focused its attention on physical causes, successful products, or human intentions, but it relates time to eternity. With each subsequent tick of the clock, another second of our lives has passed. How much time is left for us and what is the meaning of our short life? Such questions arise within a religious context. The meaning and significance of life has to do with the greatest whole that we know—and that we usually call God. The Bible says it very aptly: "In Him we live, move and have our being" (Acts 17:28). God is the space in which we, and everything around us, move and exist. In other words: God is the ground and meaning of our existence.

### *Searching to find*

So far, I have indicated four ways in which we can approach things in our lives. There are more to think of—such as the perspective of emotions, or of moral values—but I have chosen these four perspectives because they will play a role later. We have here, so to speak, four guidelines: look at things one way or another! Look for the cause, but also look for the function, or for the purpose, and don't forget to look for the ground.

What can we learn from this? First of all, this: when we seek a purpose, we should not come up with something else. Also, questions about intentions cannot be answered by answers about causes. And again, anyone looking for a purpose should not come up with a function or cause. People who still keep doing so are guilty of "bilocation" and thus commit a logical blunder.

Let me give you an example. Winking at someone is different from blinking your eyes. If someone asks me why I blink my eyes, it is a question about causes. So I can give a story in terms of causality—along the lines of: a speck in the eye, followed by impulses in nerves and muscle contractions in the eyelids. But winking is different from blinking my eyes, although they are barely distinguishable from the outside. The "why" of a wink therefore produc-

## Chapter 2: The World as Clockwork

es another story, namely a story about certain intentions that can be wrapped up in a wink. Both questions have their own answers, and therefore their own raison d'être. You may ask one or the other, but don't confuse them.

### *No hidden causes*

What is *not* allowed according to the example above is mixing up causes and intentions. The intention behind a wink is of a different order than a speck in the eye, which triggers a causal mechanism. An intention is not a disguised cause; it is not a link in a causal chain of physico-chemical processes in the body. If that is true, then biologists or physicians, looking for physico-chemical causes in the body, can never stumble upon intentions during their physico-chemical research. Anyone who wants to find out intentions must come in from a different angle.

In other words: intentions have to do with the "mind," causes with the "body." The mind can give purpose to a certain pattern of causes, but the mind itself is not a link in a causal process. As long as we continue to look through causality glasses—that is, from the point of view of causes and effects—there is no "free will," and therefore no intention, to be discovered in the brain. Intentions come from the mind, and unlike the brain, the mind is not a body part.

A simple example, once given by the philosopher Gilbert Ryle, may clarify this distinction. If you look at a football game causally, you see movements with the ball: the ball is passed from one player to the other. Suddenly someone says that the game is dominated by "team spirit." This does not mean that a special movement with the ball can suddenly be seen. No, the game is viewed from a different angle. Team spirit is not an exceptional causal link in the whole of movements on the field. Team spirit is of a different order and has to do with the overall pattern of actions with the ball. Anyone who considers "team spirit" as a special "kick of the ball" is committing a logical blunder.

Perhaps one more example: if you only focus on the individual dots in a newspaper photo of a famous person, you will never be able to discover a face. Faces are of a different order than the dots that compose them. There is no face hidden between the dots of the newspaper photo. A face is not a special dot among other dots, but it is part of a certain pattern of dots.

## Body and soul

Could something similar also be said about the relationship between mind and body? Everything indicates that our mind is not a part of the body and is not hidden between the processes of our body. Having intentions has something to do with the mind, so it is not a disguised cause that occupies a hidden place in the midst of a series of physico-chemical processes in the body. So what then is an intention? An intention is precisely the set of actions showing a certain pattern; it is different from the actions that show the intention.

## Chapter 2: The World as Clockwork

Anyone who hopes to discover a special link in the midst of the causal processes of the brain—something like a free will—would be on the wrong track according to this reasoning. Asking yourself how an intention in the mind triggers a neurophysiological mechanism in the brain is as strange as asking yourself how a computer program manages to get chips to solve a mathematical equation. That question is just wrong. Something similar can be said to someone who traces the free will back to a neurophysiological mechanism.

The body works with causes and effects, but not so the mind. Physical activity causes hunger, but the thought of "two times two" does not cause the thought of "four." If it did, we could skip many years of education, during which we must learn how to apply mathematical and logical rules. Mind and body are incomparable entities. Apples, for example, can be compared with pears, but certainly not with numbers. Finding the square root of a number cannot be done with a laboratory test. Finding the sum of the three angles of a triangle cannot be done in the lab either. The mind eludes the search of scientists.

Can we perhaps extend this line of thinking about the relationship between mind and body to the relationship between God and the world? Probably so! Talking about mind and spirit is in some ways like talking about God. Talking about the mind is done from the point of view of intentionality and rationality. That is the framework within which we look for intentions and reasons. Analogously, speaking about God is always done from the point of view of religiosity. That is the framework within which we search for ground and meaning. No wonder then that God cannot be "trapped" by some ingenious kind of experiment, for God is not a material entity.

God is the ground and meaning of our existence—God is the space in which we move and live and have our being. But if God is the *ground* of my existence, then He cannot also be the *cause* of my existence. Of course, I can say in a casual way that God is also the "cause" of my existence, but do not take this in terms of a causal mechanism. Let's keep our terminology straight to avoid confusion. God is the *ground* of my existence rather than the *cause* of my existence. God is not a hidden cause. I must look to my parents for the cause of my existence, but God is the ground of my existence. Also, the cause of my healing can be found in doctors and medicines, but that does not exclude that God is the basis of it all. Whoever seeks a cause for all of this should not come forth with God.

So, no matter how hard we try, we can never encounter God as a causal link in the great world events. He is not the first cause that unleashed the Big Bang of our universe in a sequence of events; nor is He the umptieth cause that set into motion the evolution of life on earth, and subsequently the evolution of humanity. Speaking of God as the "First Cause" is very legitimate, of course, but should not be understood in the sense of "first" in a temporal sequence of causes, but rather as "first" in the sense of being the ground of all that is. Without the "First Cause" there could not be any causes we are familiar with.

In classical theology it is said that God is transcendent—that is, He is in "heaven" and transcends the closed causality of "earth." He

## Chapter 2: The World as Clockwork

is not an immanent link in the "earthly" whole. And if there is some missing link in our scientific explanations, it can never be God. Those who look for causes—and every natural scientist does so by definition—can never come across God. The blank spots on our scientific maps must therefore never be filled with God. God is only "visible" to those who are looking for ground and meaning. God is not a "stop gap" (*Lückenbüszer*), as the theologian Dietrich Bonhoeffer once put it.

In short: there is no God to be discovered in the world as long as we keep looking through glasses of causality—that is, from the point of view of causes and effects. God can give a meaning or ground to a world of causes, functions, and intentions, but God Himself is not a link in this set of causes, functions, and intentions. The basis of our existence is precisely the framework within which all processes of the world find their place and can take place.

### *God the clockmaker?*

According to this line of thought, heaven and earth must be of a different order and dimension; heaven and earth each have their own perspective so to speak. Whereas causes, functions, and intentions have to do with the earth, ground and sense have their "seat" in heaven.

Yet it remains an age-old temptation to approach God through the lens of causality. So-called Deism sees the world as a well-running clockwork, whose clockmaker is God. Once, "in the beginning," God is said to have created the world and set it in motion, after which He disconnected Himself from His creation, retreating in His majesty.

One of the consequences of believing in a God who set the world in motion is making the world look like a mechanical clockwork. It leads to a form of *determinism.* Scientists tend to agree with determinism—perhaps even understandably so. Science would not be possible if it did not assume that like causes have like effects and that the future depends on the past. If like causes do indeed always produce like effects, then the past must determine the future entirely. Today's weather, for instance, depends on what happened yesterday, and so will tomorrow's weather depend on today's. The French astronomer Pierre Simon Laplace probably worded this in all its consequences: "We may regard the present state of the universe as the effect of its past and as the cause of its future." In this view, we would all be in the iron grip of determinism—the future completely determined by the past.

There are serious philosophical reasons to reject extreme determinism. Believing in complete determinism takes us into a vicious circle: if there is only room for determined things in determinism, then the claim of determinism must be one of those determined things, too—so it is not something I can freely choose as my worldview. Put differently, why debate human free will if all the participants in the debate are already determined to either believe in it or not? Apparently, the claim of rigid determinism just destroys itself. Even in a world ruled by the law of cause and effect, there is also our own free ability to be the cause of new events. In other words, people who defend complete determinism want us to *choose* their conviction that human beings *cannot* choose. That is a case of contradiction—if it is true, it becomes false. Hence, there is no reason to defend determinism rigorously. For the same reason, we must also reject Deism with its implications of complete determinism.

There are also serious objections to Deism from a theological point of view. In Deism, God is downgraded to a demiurge who, with His "kick-off," started the game of creation. Although He is not one of the players, He is nevertheless a causal link—albeit the first—in the midst of all the movements happening "on the field." But that is contrary to His transcendence; after all, God is in heaven and not on earth. Moreover, the Bible shows God as being involved in the events "on the field" not only at the beginning of the game—as Deism claims—but also during the game. God cannot be reduced to a cause, for God's Divinity and earthly causality are very different from each other.

**A harmonious creation**

Yet, there is still the possibility left to relate God with functionality. Biologists are known to put on the glasses of functionality. Living nature shows us many examples of successful results: males and females mate so they can have offspring; flowers are beautifully decorated and thus attract insects; blood circulates and can thus distribute oxygen and nutrition; many organisms wear protective colors and thus escape the attention of their predators. We could go on with more and more examples. The message is always the same: living nature is extremely beautiful and functional. Everything serves a purpose and produces successful results—or so the functionalists claim.

Theology has also succumbed to this kind of functionalism from time to time. Two centuries ago, William Paley formulated his "natural theology." In it, he claims that the world is based on a perfect divine design, and that everything in living nature has a function. Paley saw God as the great engineer who designed our world completely in a purposeful and harmonious way. The orderly design in creation would indisputably point to a Creator. In this view, the functionality in living nature is said to be due to the rationality of God.

However, the facts are different. Living nature is not only a romantic idyll, but also the spectacle of a cruel struggle to the death. The forces of nature can have constructive as well as destructive consequences. Volcanos can create beautiful islands and mountains but also devastating destruction. The power of growth makes flowers and babies develop into something beautiful, but it also makes tumors get bigger and bigger. Weather patterns may be the cause of a gentle breeze as well as a destructive tornado. A while ago, biologists began to wonder whether everything in living nature has a function. Many chance factors also play a role in the evolutionary process, which means that some developments are not functional, but at best neutral.

Yet, it can still make sense to keep looking for functionality—biologists can't help doing it. Of course, one can then ask oneself what the cause is behind the existence of all kinds of successful

## Chapter 2: The World as Clockwork

products in living nature. In other words: what is the causality behind the functionality? At that point, the functionality glasses are removed and replaced by the causality glasses. The question then is: how did successful adaptations in living nature arise and spread? Paley's natural theology answers this question by suggesting God's creation as the cause of this; God is the one who created all living things in such a rational way that they are functional.

In a theological sense, such an answer is not quite pure: it deprives God of His transcendence and makes Him an immanent link in evolution. But such a position is also philosophically unjustifiable: those who look for causes should not cite disguised causes—and "creation by God" is such a disguised cause. St. Augustine was already aware of the fact that "creation" is not a causal event in time, for the world was not made "in time" but "together with time," as he put it. The rationality of God can never be the cause of functionality in nature. Instead, it is the very ground of all there is.

Still, it would take some time before biologists began to seriously look for real causes in time. Charles Darwin was one of the first to do so; he therefore spoke of evolution, not in terms of creation. He came up with the hypothesis of an evolutionary mechanism based on natural selection. Natural selection works through survival and reproduction by promoting what is successful, while suppressing what is unsuccessful. The cause of functionality in nature must therefore not be sought in the rationality of God, according to Darwin, but in the causal process of natural selection.

So much for the perspectives of causality and functionality. They do little to help us relate the world to its Creator, God. So, we are left with the irreducible question: what is the ground and meaning of this world and of our existence in it? Certainly, an interesting question with interesting answers. But we should not expect answers to come from the natural sciences. Those answers must come from a different angle.

# Chapter 3

## Does Mother Earth have a Heartbeat?

Organisms are masters when it comes to the art of steering. They do not only steer through their body, but also within their body. From minute to minute, all kinds of processes are being checked and adjusted inside the body so as to keep important values in the body within the correct limits. Just think of features such as heart rate, body temperature, or glucose level in the blood. As soon as the results become too high or too low, action is taken and, as a result, these body values fluctuate within fairly narrow limits. This is also called self-regulation, which means that deviations within the system are controlled according to some preset norms or standards.

The mechanism of self-regulation is based on the fact that cause and effect interact with each other so that the effect in turn influences its own cause. In essence, such a cause-and-effect is linked back to its beginning through a certain norm: while the cause has a reinforcing influence on the effect, the effect in turn has an inhibiting influence on its own cause once the norm has been surpassed. That is why this is also known as a "feedback" mechanism. It is thanks to this mechanism that self-regulation is possible. This does not mean, though, that any form of feedback automatically leads to self-regulation; there must also be a standard built in.

### Masters of seamanship

There are many examples of self-regulation based on feedback mechanisms. Take these two processes for example: 1. When the body temperature increases, the radiation of the body increases as a result, which causes the body temperature to decrease again. 2. When the concentration of glucose in the blood rises, the concentration of insulin also increases as a result, causing the concentration of glucose to decrease in turn. These are fine examples of self-regulation. Somehow, our bodies know what helmsmanship is.

So far, these are obvious examples. At first glance, the following innocent example also belongs in this series: 3. When the number of prey increases, the number of predators increases as a result, so that the number of prey decreases. This seems to be another example of steering skills found in nature. But some caution is needed.

In a strict sense, this is indeed a feedback mechanism, because the effect (more predators) has an inhibiting influence on its own cause (thus fewer prey animals). That's why terms such as "self-regulation" and "natural balance" are often used in this context. Mother Earth supposedly balances the number of predators and prey well, as befits a good mother. Isn't that right thinking?

I do not think so. In order to speak of regulation or equilibrium, not only is a feedback mechanism required, but also a pre-set norm that has been adjusted to the correct value—supposedly by the evolutionary process of natural selection. The balance of prey animals and predators would then have to fluctuate, as it were, around a certain standard value. The question now is: what is that

# Chapter 3: Does Mother Earth have a Heartbeat?

"natural norm" of this process? Because if there is no norm involved, then there can be no question of a natural equilibrium. So, there is a lot at stake.

## *A natural norm?*

It is tempting to look for this so-called "natural standard" in the "carrying capacity" of the environment. After all, the environment can no longer accommodate animals above a certain number. Above that number, the earth, or the living environment, is simply full and the carrying capacity has therefore been exceeded. But beware: the concept of "carrying capacity" does not relate to "balance" but to "over-capacity." In other words, the carrying capacity is not a standard value but a maximum value. And that is why we need to look for the desired standard elsewhere.

Perhaps the average condition or the average number of prey or predators is a good standard? But we must be careful about proclaiming the average as the norm—if only because the environment is constantly changing, and with it the so-called standard as well. Feedback between the number of prey animals and the number of predators simply cannot have a regulatory outcome, because such a feedback mechanism would be "limping behind" after at least one generation. Just think of this: because the number of prey animals has also changed in the next generation(s), the number of predators that has changed in the meantime no longer matches the number of prey animals that existed at that time. If there were a fixed relationship between the two quantities, the consequence would be that the fluctuations in numbers should gradually become larger instead of smaller!

Apparently, there is no question of self-regulation or equilibrium here, and therefore it seems better to avoid the term "natural equilibrium" from now on. It would be wiser to replace this misleading term with the more neutral notion of *stability*. Stability only means that the fluctuation of numbers of predators and prey remains within narrow limits. It is left open whether the average level of the fluctuations "should" be at a certain height and whether that level is the result of control and regulation. The average number of prey or predators does not act as a norm but is merely seen as the result of a (large) number of processes that are largely independent of and sometimes opposed to each other. Stability in nature therefore has everything to do with probabilities and probability distributions, rather than with a fixed, ideal standard. Predators and prey keep each other in check, but not in balance. Their numbers are not under control nut under constant pressure.

### Risk spreading

Why then is it that ecosystems and biotic communities usually do exhibit a certain degree of stability, even though there is no question of self-regulation or natural equilibrium? In other words: how is stability possible without self-regulation? The answer roughly goes as follows. A strong stability of an ecosystem is associated with a great diversity of species within that system. The more species a system harbors, the more stable the system is supposed to be. This is not a question of regulation based on an ideal standard, but instead has everything to do with probabilities and probability distributions. The more complex an ecosystem is, the better its "risk spread" is, thus making it more resistant to fluctuations. So what we see here is not a matter of regulating a balance, but of spreading the risk.

A large diversity of species therefore has everything to do with a good risk spread. An ecosystem is less sensitive to fluctuations if it contains more species with more interconnections. Just think of a rich system such as the tropical rainforest, which is generally quite resistant to disturbances. On the other hand, a poor system with one dominant species such as an agricultural area is hardly able to withstand pests. A simple story, you would say.

But it is not that simple, because in a system with a richly branched network of species, disruptive influences may not provoke large fluctuations, but possibly ripples that are so numerous and widespread that the whole can nevertheless be considerably disturbed. In that sense, even a tropical rainforest is quite vulnerable. It just depends on how stability is defined: based on the size of the ripples, regardless of how *long* they last, or based on the duration of the ripples, regardless of how *large* they are? You can see how we can easily lead each other astray with vague terms.

### *A super organism*

In short, self-regulation is something that takes place mainly *within* an organism. But whether it also plays a role *between* organisms is very doubtful. Some biologists and geologists have found a solution to this. They have sought refuge in the *Gaia* hypothesis of the British biologist James Lovelock. According to this hypothesis, our earth would be some kind of super-organism called Gaia, and that super-organism, Lovelock claims, is maintained by various feedback mechanisms between biosphere, atmosphere, and lithosphere—just like other organisms are.

That is quite something. We already knew that life has changed the appearance of Planet Earth and that the earth has modeled the forms of life it harbors. But the Gaia hypothesis goes much further. It proclaims an active regulation in nature—that is, life keeps the earth livable, according to a certain standard of viability. Life creates a natural equilibrium around itself, and as a result, life has so far been able to withstand enormous disturbances, such as the impact of colossal meteorites, or the changing heat radiation of the sun. And by the same principle, life should still be able to withstand the threat of increasing global warming.

In short, Lovelock sees the earth as a colossal "organism" with a set of built-in norms that allow for self-regulation. In other words, the earth would react to disturbances in nature similar to the way an organism reacts to fluctuations in its blood. Consequently, there must be active regulation, making the living world in control of the inanimate world. Take the quite stable temperature of the earth, for example. Despite the fact that the heat from the sun has increased by more than twenty-five percent since the beginning of life, the earth has remained fairly constant in temperature. It seems as if "Mother Earth" maintains its own body temperature. The next

# Chapter 3: Does Mother Earth have a Heartbeat?

question then is: which biological control mechanism lies behind this?

In response to this question, Lovelock developed the *Daisyworld* model. Suppose white and black daisies grow on the earth. With little sunlight, the black daisies grow better than the white daisies because they absorb the heat better. As a result, more black daisies will appear, and the temperature will gradually rise. As soon as the temperature has risen enough, the white daisies have now gained an advantage, because they reflect solar heat better. The white daisies now take over the place of the black ones, and the surface of the earth cools down again. As a result of these feedbacks, the temperature continues to fluctuate within certain limits.

Do these feedback mechanisms really result in a stabilizing effect? Not quite. Feedback mechanisms only have a regulatory effect if they are linked to a pre-set norm or standard. Where is the "thermostat" of this temperature regulation located? It has been suggested that certain types of phytoplankton act as a thermostat: as the temperature in the atmosphere rises, they increase in size and therefore produce more of a certain gas (dimethyl sulfide), which stimulates cloud formation in the atmosphere; as a result, cooling would occur again, followed by reduced phytoplankton production, and so on.

### *Where is the proof?*

Is the Gaia hypothesis testable asis required for a good scientific hypothesis? Testability means, among other things, that the hypothesis can also be refuted. Can the hypothesis of global temperature regulation be refuted? Yes, it can be refuted and has been refuted in fact. Research has shown that the size of the plankton population does in fact *not* respond directly to temperature changes, as you would expect from a good thermostat. There appear to be many other factors at play. Apparently, here too we are not dealing with a regulated balance, but rather with a spread of risk. And that means the Gaia hypothesis doesn't find enough empirical support. If there is a global thermostat based on phytoplankton, it doesn't work very well. Isn't that a fatal blow for the Gaia hypothesis?

No, the proponents of the Gaia hypothesis are not so easily defeated. They are ready with many other regulation mechanisms. The following is one of them. Many trees appear to release enough of the (highly flammable) isoprene to sustain an incipient forest fire. In addition, the oxygen content in the atmosphere is maintained at 21%, partly due to the presence of methane. Thanks to a combination of these factors, forests have a certain number of wildfires, enough to rejuvenate themselves and thus ensure their own survival. Isn't that fantastic? Yes, the Gaia hypothesis is a comforting thought. Mother Earth is skilled in the art of steering and watches over us day and night.

But, again, is the hypothesis correct? In the case of forest fires, it is especially difficult to falsify the hypothesis, because the model has so many interrelationships built into it that there are all kinds of escape routes. Moreover, the Gaia hypothesis can be watered down to the point where not much of its original strength remains.

# Chapter 3: Does Mother Earth have a Heartbeat?

It then boils down to the safe proposition that the environment influences life, and that life in turn acts on the environment. But that's quite obvious. Another way out is the claim that the Gaia hypothesis merely confirms that life has been able to survive on Planet Earth so far. But we already knew that, even without the Gaia hypothesis. If life on earth hadn't been able to survive, there wouldn't be humans to come up with the Gaia hypothesis right now anyway!

## At the cradle of Gaia

However, the special thing about the Gaia hypothesis is that it goes much farther: planet earth is supposed to be a living superorganism that maintains an optimal physico-chemical environment for life. Mother Earth, so to speak, keeps her finger on the pulse. This means not only that life is adapted to the environment, but also conversely that the environment would be adapted to life. "Adaptation" is the key word here. Now, adaptation is all about evolution; organisms have become self-regulating by undergoing the test of natural selection. This capacity has been developed through "trial and error" and that's how the norm has been adjusted to the correct value.

But if the earth really, like an organism, were to form a self-regulating system, how did the earth become self-regulating? Now the question we should ask the Gaia hypothesis is how such a self-regulating superorganism as Gaia ever came to be. How did Gaia ever learn the art of helmsmanship?

Obviously, there can be no question of natural selection, as there is only one Gaia system—and thus natural selection of different Gaia variants is excluded. Temperature regulation is apparently not an "ability" of the earth that has arisen through natural selection and adaptation. But then it can't be much more than a "coincidental circumstance" that Planet Earth has been equipped with. But whether that fact is also a guarantee for the future is doubtful. Gaia hosts a network of interactions, and thanks to their multiplicity these have at best a stabilizing effect, but they do not have a regulatory effect oscillating around a pre-set norm. In short, the so-called natural balance of Gaia turns out to be again a matter of "ordinary" risk spreading. We are dealing more with a stabilizing effect than a regulatory effect.

So, what remains of the Gaia hypothesis? I fear little more than an unfounded but perhaps comforting thought: Mother Earth watches over us and keeps monitoring her heartbeat. A thought that is not based on factual material but was happily embraced in the 1970s by New Age environmentalists. Giving "Mother Earth" a nickname, Gaia, does not make her more realistic. She does not have a heartbeat. She does not even qualify to be called that way. But, you might ask, is the thought of Mother Earth not quite comforting? I have my doubts. Frankly, I feel much safer with the thought that there is Someone Else than "Mother Earth" watching over us.

# Chapter 4

## The Secret of Life

Today, almost every student in high school knows that people are made of cells, that cells are built up from molecules, and that molecules are made up of atoms. This knowledge is the result of a long, respectable, and successful tradition in the natural sciences: take everything apart, down to the smallest parts. Dissecting, unraveling, and analyzing, that's what science is good at.

Ultimately, it seems, every researcher should be concerned only with the smallest of the smallest. And those are atoms—or rather quarks, because those atoms are built up from them. And we can safely leave everything else, such as cells and organisms, behind us. That is pretty much the program and ideal of what is called *reductionism*: small is beautiful!

*Extremists*

Supporters of reductionism claim that the properties of an assembly of parts always depend entirely on the properties of those individual parts. All that really matters are the smallest parts. Reductionists therefore believe that the "higher" levels of reality—such as the level of organisms or the level of cells—can be completely reduced to the lowest level—that of atoms and quarks without losing any significant information. To put it with an example: behind every individual there is nothing but quarks.

And yet there are still scientists who are not concerned with atoms and molecules, but with cells, or even with organisms, not to mention populations or ecosystems. Are they a dying race, the last of the biologists? Or are the opponents of reductionism, assembled under the name *holism*, right after all when they say that a cell is *more* than the sum of its constituent molecules—or that an organism is more than the sum of its constituent cells? The clash between reductionism and holism has given the above problem a more general scope: is the whole equal to the sum of its constituent parts, or is the whole greater than the parts? Aristotle himself used to say, "The whole is more than the sum of its parts." The answer to this question is not as simple as the "extremists" tend to think. Why not?

The opposition between these two ideologies, reductionism and holism, became acute during the history of embryology at the turn of the previous century. The discussion was triggered by the first experiments with embryos during their two or four cell stage. It turned out that by killing (puncturing) one of the four cells of a frog embryo, the further development of the embryo did not proceed at all, or it proceeded in an abnormal way. That was good news for reductionists. You see, they said, the parts determine the whole all by themselves. Their explanation was simple: one cell out of four can only contain the hereditary factors for a quarter of an organism. If a part no longer functions properly, then the whole is affected. According to this view, it is the parts that determine the whole.

But other experiments were done, too, and they were grist to the mill of the holists. As soon as the two or four cells of a sea urchin egg were separated by shaking, it turned out that the isolated cells could still go through normal embryonic development. You see, the holists said, the whole determines the parts. After all, this experiment clearly shows that the control of embryonic development comes from something more than the individual cell. "Something" tells the detached cell that it must grow into a new organism all by itself. That must be a signal that comes from the organism as a whole. Some holists even spoke of an immaterial, vital principle of life that would direct, regulate, and attune life's phenomena to an ultimate goal—which is why this view is sometimes called *vitalism*

**A timeless conflict**

Who is right? On one side are the reductionists, the defenders of the parts of the whole. They claim, for instance, that every subsequent step of development is entirely dependent on DNA. This point of view is currently pre-eminently held by molecular biologists. They form a growing contingent of biologists who have unraveled the code of heredity ever further. They have discovered that the parts of the code are much more varied and complex than is often thought. Thus, in addition to normal genes, there are also *regulatory* DNA sections that turn ordinary genes on and off. Furthermore, DNA segments can change position and thus influence the activity of other genes. Finally, it has been found that DNA segments can be rearranged in such a way that millions of different proteins (in case of antibodies) can be made from less than a thousand genes. In short, the whole consists only of parts, even though they are "ingeniously" put together.

On the other side of the conflict are the holists, the defenders of the whole behind the parts. Nowadays the term holism has almost become a dirty word, as it covers so many different things. But that's not quite right. Holists emphasize the importance of the whole; they will place the "initiative" for a new developmental step rather or exclusively with the entire structure or organization of the organism. After all, the DNA system depends on signals from surrounding cells and organs; these tell the DNA which genes to activate at what time. There is a gradient across the entire organism, for example from head to tail, which tells the individual cells where they are located inside the embryo and which genes they must "activate." Without the whole, the parts cannot do much. Some holists even claim that the genetic code is controlled exclusively from the outside.

Perhaps we can learn something from this historic conflict. It has become apparent that embryology, and probably all of biology, cannot be squeezed into the straitjacket of either reductionism or holism. On the one hand, DNA turns out to be a regulatory system that is much less rigid than is often thought. The reductionist approach has made that clear to us. On the other hand, the DNA system is not completely autonomous, but receives external signals from the surrounding cell plasma and surrounding cells, tissues, and organs, which activate certain genes at the right place and at the right time. We mainly owe this insight to the holistic approach.

**No secret powers**

Now we can go back to the original question: is the whole after all more than the sum of its constituent parts? No, at least not in the sense that additional parts would be necessary. Otherwise, we will be tempted to add an extra part to the constituent parts—something like an immaterial force or principle. Vitalists tend to do that; for years they have been looking for something extra that holds the parts together and steers them. They introduced non-physical elements, some vague, elusive principles that a scientist can do little or nothing with. Like this: a cell is a collection of molecules plus "life"; and an organism is a collection of cells plus a "soul." But how do you scientifically investigate something as elusive as "life" or "soul," or any other non-physical forces?

For centuries we have heard claims that life forces are capable of making living beings out of inanimate substances. Aristotle had already argued that sunlight, rotting flesh, and even mud are capable of producing living beings. It was not until the 17$^{th}$ century that some biologists began to question this. If life "suddenly" arises somewhere, then—as we now know—animal eggs or traces of bacteria must have been present there, although these can sometimes easily be hidden from our sight. Eventually, it was Louis Pasteur's experimental work that proved conclusively that where there are no organisms, no organisms can arise—or at least not through something as elusive as "life forces." Conclusion: the whole is nothing more than the sum of its parts.

And yet, in a certain sense, the whole is also more than the sum of its constituent parts. Nowadays everyone knows that if you put the right molecules together in a test tube, you don't get a cell; and if you put cells together, you get at best a tissue culture, but no organism. Apparently, a cell is "more" than a collection of molecules, and an organism is "more" than a clump of cells. But what is that "more" then?

## The secret of life

Perhaps an example can help us. Take the newspaper photo of a face again. By paying close attention to the individual dots one can never perceive a face. Why not? Is the face perhaps "more" than the individual dots? The answer is no, because all we need to see a face are the dots printed by the printer. And yet the face is "more" than the dots, for the face is not a property of the dots as such, but of the way in which all the dots are arranged together. This "more" is to be found in the structure and in the mutual relationships between the parts. But relationships are something other than properties. Two people with good qualities don't always make a good relationship, and you don't automatically create the best football team by putting the best football players together. Apparently, the properties of an assembly cannot be traced back to the properties of the parts, but to the relationships that the parts have with each other.

This can also be said about a cell or an organism. They have a certain, almost unique structure of parts, which together stand in a certain relationship to each other. So, it is extremely important how the molecules in a cell, or the cells in an organism, are arranged. Within such a structure, something very curious can happen: a certain arrangement of molecules makes for a cell. And within that assembly, parts may be replaced without affecting the whole, so long as the structure remains unaffected. The existence and functioning of a body cell, for example, is a consequence not only of the existence and functioning of its cell parts, but also of the existence and functioning of the entire organism, of which this cell is a part.

Now it is also clear why the contribution of physicists and chemists to biology is so great. It is indeed important to examine the building blocks of life in greater detail. That is the way "down"—down to the smallest details—and that road is being followed by (bio-)physicists and (bio-)chemists. But on the other hand, we will always need "real" biologists, because we should never lose sight of the structure of larger wholes; and that is the way "up." This latter pathway leads to "higher" entities: cells, organisms, populations, and ecosystems.

### Two-way traffic

This two-way traffic—between "bottom-up" and "bottom-down"—will probably be of all times, although the way "down" has been particularly popular lately. And that in turn has to do with the many successes that have been achieved with the "bottom-down" approach, because cut-apart pieces are simply easier to investigate than large, complex systems. Blood in a test tube (*in vitro*) is easier to test than blood in the body (*in vivo*). But let's not forget that blood in a test tube is already "dead" blood.

Because the way "down" has been so successful, we are all familiar with the popular slogan "The secret of life is in the DNA." During their "descent," biochemists and molecular biologists have discovered that many of an organism's appearances can be traced back to minute details of the DNA molecule. Isn't it amazing what unraveling can yield!

But we forget so easily that on the way down we are going in the direction opposite of the way up. While going "down," we quickly lose sight of the fact that this "glorified" DNA molecule is merely part of a much larger whole. The DNA is just one of many links in a complex process of information transfer within the cell. To make the right proteins in a cell, not only is DNA needed, but also many other components such as ribosomes, RNA, and enzymes. And moreover, we easily forget that this process in turn must receive steering signals from the organism as a whole, and even from its surroundings. In other words, DNA (in a test tube, for example) cannot do anything as long as it is not part of a larger system.

Anyone who claims that the secret of an organism lies in its DNA could therefore just as rightly launch the reversed slogan: "The secret of the DNA lies in the organism." Expressed in an image: it depends on from which side of the telescope you are looking—from the side that enlarges or from the side that reduces. In analogy, with one side, one can see a face, with the other side only the dots that make up the face. Unfortunately, the users of the "scientific telescope" tend to forget that there is in fact another side to the dots.

# Chapter 5

## The Origin of Life

Where and when did life begin? It's a question that is probably as old as humanity itself. But the correct answer to that question is far from ancient. If you have been properly educated—but only recently—you will at least know where your own life started. That was when a fertilized egg came into being. Your life indeed began with that fertilized egg. But not so life itself, because a fertilized egg requires existing cells—a fusion of an egg cell and a sperm cell—and these in turn arise from other cells. So the question is then whether there have ever been cells without previous cells, and if so, when and how did they originate.

As you can see, the question of the origin of life has been reduced to the question of the origin of the cell. Today "life" is identical with a "cell." The cell is the modern, basic scientific unit of life. Profound questions about life—questions that philosophical and theological thinkers like to ask—are narrowed down by modern biologists to technical questions about the cell, that is, about a packet of DNA or RNA surrounded by a cell membrane and equipped with its own metabolic mechanism. Biologists only speak of life when something consists of at least one cell. Because a virus only contains DNA or RNA and is otherwise dependent on existing cells, we cannot speak of "life" in the case of a virus. It's that simple.

### The basic unit of life

Anyone who asks about the beginning of life would therefore be better off asking about the beginning of the cell. So our question should really be: are there cells that have not evolved from other cells? And if so, did that happen a long time ago or is it still happening?

Even before people knew that life is related to cells, these questions were no different. Some said that the chain of life started only once; others claimed that life could re-emerge at any moment. The latter view has long been very prevalent, and therefore I would like to show you what has become of that view.

If you don't have a microscope and can't see cells, it's easy to get the impression that life is constantly springing up out of "nowhere" around you. The great philosopher (and biologist!) Aristotle saw something like this happening all around him, precisely because he was such a keen observer. He rightly spoke of the fact that life arises not only from seeds and eggs, but also from carcasses or even from mud—from non-living things, that is. Just as a fire in a haystack can start spontaneously (that is, without matches or the like), so can life repeatedly and "spontaneously" arise even in cadavers (that is, without seeds or eggs). That is why Aristotle spoke of *generatio spontanea*.

Aristotle's explanation of this "fact" was, of course, consistent with the rest of his philosophical system. In addition to a source of nutrition, life also requires a life force, an active principle. Upon fertilization, the egg provides the source of nutrition, but the egg

## Chapter 5: The Origin of Life

can only develop into an adult organism thanks to a life force (which is supposed to come from the sperm). Well, something similar applies to "spontaneous generation." When a life force (from sunlight, for example) enters the breeding ground of rotting flesh, new life can spontaneously arise. And so it is that mice can spontaneously arise in Cologne pots of grain, or that swarms of flies can develop on cadavers of cattle. This is not a matter of popular belief, according to Aristotle, but of common sense and sober observation,

How is it possible that such an obvious explanation did not make it into the 20$^{th}$ century? We owe this to the triumphal advancement of science, as many natural scientists will tell you these days. For example, the late Belgian Nobel Prize winner Christian De Duve says that it was Louis Pasteur who gave the death blow to the theory of spontaneous generation. In the first part of this chapter, I want to show you how lethal this blow was and in the second part what happened afterwards.

### *The fatal blow*

There is a rather recent American biology textbook for high schools that teaches students how to conduct scientific research. And that is exactly what the success story of Louis Pasteur is used for! Pasteur had let the scientific method triumph over the old popular belief of "spontaneous generation," such is the message of that book. So, we should be able to learn something from that.

To begin with, the said book puts some question marks on the "facts" that Aristotle had to report. Of course, those were not facts but interpretations, says our book. But twenty centuries later that is very easy to say. What biology textbooks today present in terms of facts, could also be time-bound interpretations. Biology books are not only constantly being revised and expanded, but also rewritten in detail. After all, there is no information without interpretation, as we discuss in the last chapter.

Aristotle's "facts" have long remained commonplace. In the 17th century, the Jesuit Athanasius Kircher (1602-1680) was still a convinced supporter of the theory of spontaneous generation. He was able to show that "worms" can develop from dead flies on honey water. The Belgian scientist Jean Baptiste Van Helmont (1580-1644) was also able to do something like this. His recipe for "producing" mice was to add grains of wheat to a dirty shirt. Young mice emerge spontaneously after 21 days, he noticed. The grain provides the source of nutrition, and the sweat of the dirty shirt must have provided the life force—neatly according to theory.

Although Van Helmont is known for his good research into the metabolism of plants, he has made some mistakes here, according to the textbook mentioned above. The student is asked which control experiment Van Helmont should have carried out. The correct answer will pose few problems for most students of this day and age. Of course, the alternative hypothesis—namely that young mice do not arise spontaneously but arise from parents—should have been ruled out by closing off the shirt and the grain from the outside world. That is called a control test. How could Van Helmont have forgotten such a thing! The lesson is clear: it is scary so see what prejudices can do to science.

Fortunately, the Italian researcher Francesco Redi (1628-1698) knew better. He was strongly inspired by William Harvey's thesis that all life comes from an egg (*omne vivum ex ovo*). That test was therefore explicitly built into his experiments. He used open and closed bottles with meat in them. He found out that "worms" (fly larvae) only developed in the open bottles, but not in the sealed bottles. Wasn't that a final blow to the theory of spontaneous generation?

No, the "facts" are not that simple. Closing bottles is not only a matter of allowing eggs to enter, but also of allowing fresh air to come in—and that air may contain the life principle so necessary for spontaneous generation. In short, Redi's ideal experiment had one thing "under control," but not the other. That is why he then repeated his experiments with fine gauze. He found no life under the gauze, at least no "worms." So the air did not help!

However, the matter became more complicated when Anthonie van Leeuwenhoek (1632-1723) saw "small animals" under his microscope. He saw them increase in number but not in size. Isn't that some form of spontaneous generation? Perhaps life is constantly emerging, which we can only observe with the microscope. This gave the discussion a new impetus and had now moved the issue to the micro level.

In 1745, the Englishman John Needham worked with bottled fruit juices that he had heated and then sealed. Despite heating and sealing, the mixture in all bottles began to rot. This seemed to suggest that Needham had revived the spontaneous generation debate. It will also be clear that the biology textbook mentioned above now asks the students: What do you think went wrong in this experiment? The answer will probably be that the liquid was not sufficiently sterilized or that the sealing was not done under sterile conditions. But things only go wrong if you assume that rotting is supposedly not allowed to occur in sealed bottles. In the 18th century it was precisely this last assumption that was under discussion.

But not so for the Italian priest Lazzaro Spalanzani. He simply did not believe in spontaneous generation, so in 1770 he boiled his bottles of nutrient medium for an hour, after which they were hermetically sealed. No rotting was found in any of his 19 bottles. That was to be expected, according to Needham, for his opponent had mistreated each bottle by prolonged and high heat, so that the life principle "must" at least have been weakened or perhaps destroyed. Also, this time there "must" have been something wrong with the treatment, but now based on the arguments of the opposite party.

That is why Spalanzani decided to repeat his experiments, but this time with different cooking times and without hermetically sealing the bottles. If the so-called life principle is weakened by cooking, then there should be less rotting the longer the cooking time was, he reasoned. In fact, he found rotting in most bottles—just the opposite of what was claimed.

## *A decisive experiment?*

For another century the discussion went on like this—discussion is the engine of science. In the meantime, people had started to discover that air can be not only full of life force, but also full of micro-organisms. It therefore became important to investigate whether air without micro-organisms is capable of spontaneously generating life. If it does, then there must be something like a life principle in the air after all.

And so that is going to be Pasteur's decisive experiment. Louis Pasteur (1822-1895) designed a swan neck for his flasks, which allows air (and perhaps life force) to pass through but no micro-organisms (because they get "stuck" in the bend of the neck). What did he discover? In such flasks, boiled fruit juice could not show rotting, despite the free access of fresh air. And after filing off the swan's neck, rotting appeared again, despite the possibility that the so-called life force had been destroyed by cooking. Pasteur's experiment had finally decided against the theory of spontaneous generation. At last, science and its scientific method had triumphed.

But how decisive was Pasteur's experiment? Pasteur had an opponent, Felix Pouchet (1800-1872), who would later become the big loser in this "debate." He is hushed up in most books, because the battle was already settled before Pasteur's decisive experiment. What exactly was going on?

In 1861, Pasteur worked with fruit juice in gooseneck flasks. The boiling caused the air to disappear from the flask and the opening was sealed. This way the contents remained unaltered.

Then the end of the neck was broken off with heated tweezers, but in clean air—that is, in air free of microorganisms. This happened at an altitude of 2 km in the French Alps. Well, 19 out of 20 flasks showed no rotting. Pasteur concluded: there is no spontaneous generation.

In 1863, Pouchet did the same experiment, high up in the Pyrenees, but he used flasks of hay suspension instead of fruit suspension, plus a heated file instead of heated tweezers. As a result, all eight flasks showed rotting. So, Pouchet concluded that putrefaction is apparently due to spontaneous generation. Pasteur's response was: no, the air there must have been contaminated with micro-organisms, otherwise rotting cannot possibly occur. You don't expect such a yes-no game in science. That calls for a commission.

And it came. The French *Académie des Sciences* installed the committee in 1864. This committee consisted mainly of members who had already spoken out against the theory of spontaneous generation. Pouchet was willing to demonstrate that all (!) of his flasks would change if air could enter them. But the committee was already satisfied if his opponent, Pasteur, could demonstrate that some (!) flasks would not change if air could enter them. Pouchet withdrew as soon as he learned of the committee's composition and its biased demands. In fact, the outcome was already decided.

It is noteworthy that the committee never demanded to test the difference in tools (file versus tweezers) or the difference in nutrient medium (fruit versus hay). Although Pasteur complained that a file can cause more contamination than tweezers, as far as we know

# Chapter 5: The Origin of Life

it was never tested (a spicy detail: at Pasteur's request, his laboratory reports remained inaccessible for a century; but now that they are accessible, Pasteur also appears to have committed fraud). And the difference of fruit versus hay was never even brought up. Nowadays we know that hay often contains a bacterial spore that cannot be destroyed, not even by prolonged cooking. Had Pouchet continued his experiments with hay, he probably could have given the committee a big surprise. That's how it works in science.

In other words, Pasteur's experiments were not as decisive as people often think today. But Pasteur had the wind in his side; the theory of spontaneous generation began to fall apart. Besides, there was another reason for that. In the meantime, Charles Darwin had published his theory of evolution, and this theory initially had a lot of wind against it. Evolution, and certainly evolution of the living from the non-living, could only have sprung from fantasy. At last, there was now a "sound" scientist, Louis Pasteur, who could prove irrefutably that no life can arise from the inanimate. So that "proof" was actually killing two birds with one stone: the death blow for the theory of spontaneous generation, but also the death blow for the theory of evolution.

As we have already seen, the first blow was not as deadly as Pasteur had suggested. Sterilization procedures still needed to be refined. Destroying heat-resistant bacterial spores sometimes requires high temperature and high pressure, or a certain alternation of heat and cold. As late as 1910, Henry Bastian dared to support the heresy of spontaneous generation again as an explanation for the fact that some life forms may outlive heat. Facts can be as stubborn as heat-resistant bacterial spores.

The second blow was also not fatal. Charles Darwin's theory of evolution gradually began to gain traction in the world of biologists, and so people began to look for the moment of the emergence of life from the inanimate—not as a recurring event, but as a one-time process of long duration. In the second part of this chapter, I would like to devote some attention to this research.

### A primal cell?

Fossil remains that resemble bacteria have been found in rocks that are about 3,500,000,000 (3.5 billion) years old. Although our planet was formed about 4,500,000,000 (4.5 billion) years ago, we know of no fossils from those first billion years of our planet. If we accept these dates for the time being (within certain margins of error, of course) and assume that there is such a thing as evolution, we must also assume that the first cells had about 1,000,000,000 (1 billion) years to develop from components that were not yet cells. Moreover, if we may assume that there is no spontaneous generation based on life forces, we must look for something else that initiated this transition from inanimate to animate.

A "modern" biologist will then look for an energy source that is able to boost simple raw materials into larger compounds, and ultimately into a "primal cell." Nowadays, organic molecules, such as carbohydrates, nucleic acids, and proteins, are built up by cells with their own energy (obtained from food or from the sun). But those cells were not there yet! How then could complex organic molecules arise?

For the construction of organic molecules, both a carbon source (C) and a hydrogen source (H) are needed. The carbon source was probably, as with green plants today, carbon dioxide ($CO_2$), as there was enough of that in the primordial atmosphere. The hydrogen source, on the other hand, was a major problem. Although hydrogen is typically locked up in water ($H_2O$), it is an oxidized and therefore worthless form of hydrogen. Oxygen easily absorbs hydrogen and then forms water (in Greek, hydrogen means "water-former"). To release this hydrogen again, powerful reducing agents (reductors) are needed, which counteract oxygen's absorption of hydrogen. Were there such powerful reductors in the primordial atmosphere?

In many old layers of the earth, we do find formations of "layered iron." This iron may have been involved in forming powerful reductors. Under the influence of ultraviolet radiation—and there was plenty of that in the primordial atmosphere—divalent iron ($Fe^{2+}$ or Iron(II) oxide) can be converted into trivalent iron ($Fe^{3+}$ or Iron(III) oxide). During that process, an electron is released, and that electron is a powerful reductor. Normally, such electrons are immediately gobbled up by oxygen, but in the primordial atmosphere there was practically no free oxygen. Therefore, the primitive ocean was probably infused with electrons, providing a constant source to trigger chemical reactions to build simple organic molecules. Strong ultraviolet radiation ensured another constant energy supply—a modern form of spontaneous generation, so to speak.

Which were the organic molecules that played a key role in the origin of life? This question cannot be answered based on what we know about existing cells, because cells contain both nuclear acids (DNA and RNA) and proteins (structural proteins and enzyme proteins). The nuclear acids contain the genetic code for making proteins, but that code can only be read and copied using certain enzymatic proteins. Hence, without the presence of nucleic acids there are no proteins possible, but without proteins no nucleic acids are possible either. Hence, the question is: what came first, nucleic acids or proteins, the "chicken" or the "egg"?

That problem seems to have been resolved in the meantime. Nuclear acids—that is, certain RNA molecules—have been discovered, that can also function as enzymes (and could therefore replace enzyme proteins in speeding up reactions). This RNA could therefore fulfill both tasks: it can be copied, and thus store and pass on genetic information, and it can speed up reactions without the presence of enzymatic proteins. This RNA could be a good candidate for the link between the inanimate and animate. With energy from electrons and ultraviolet radiation, carbon and hydrogen could be built into organic RNA-molecules.

Instead of a spontaneous generation, we have here a "spontaneous" formation of RNA molecules. Aristotle would be proud of it. This "spontaneous" process has even been tested in the laboratory. Under experimental conditions, simple building blocks from the so-called primordial atmosphere were exposed to a strong source of energy. Indeed, components of RNA were then formed that were not present before. True, these are not yet complete RNA molecules, but that may only be a matter of time. And, also true, RNA molecules are not yet complete cells, but a lot can happen in millions of years.

There is no real scientific evidence that something like this did in fact happen, especially not when it comes to such a long historical process as the origin of life. What we have described here is a research program rather than a research result. All we can show is that it's not impossible that something went this way or that way. Science, meanwhile, continues to investigate.

# Chapter 6

## The Evolution of Life

If you don't believe in evolution, I am not going to try to change your mind. The theory of evolution still has many questions left unanswered. However, the Catholic Church does not stop you from accepting evolution as a process. Several pontiffs have expressed that view. Pope Pius XII declared, in his encyclical *Humani Generis*, that opinions favorable and unfavorable to evolution must be carefully weighed and judged. Pope John Paul II was more explicit when he said, "there are no difficulties in explaining the origin of man in regard to the body, by means of the theory of evolution. But it must be added that this hypothesis proposes only a probability, not a scientific certainty." Later on, the pontiff would add, "some new findings lead us toward the recognition of evolution as more than a hypothesis." Pope Benedict XVI said something similar, "there are many scientific tests in favor of evolution [...] But the doctrine of evolution does not answer all questions, and it does not answer above all the great philosophical question: From where does everything come?"

On the other hand, the Catholic Church does not declare evolution as a dogma that we must accept. That is beyond her expertise—she will never give scientific theories a seal of certainty. The theories of heliocentrism, the Big Bang, or evolution were never given the status of dogma. So, what follows in this chapter is not something you must accept as a Catholic. But it is good to know more about the theory of evolution, because most people you will meet nowadays believe that evolution is indeed a fact.

**The great coincidence**

In the context of evolution, *chance* is somehow the new life principle of biology. How has the concept of chance been able to penetrate so deeply into the theory of evolution? This is mainly due to Charles Darwin. Thanks to him, we have come to see this great evolutionary process—from the primitive primordial cell up to and including the current wealth of life forms, including humans—as a succession of small processes of change: from step to step. To Darwin's great credit, he reduced that complex, vast, and lengthy process of evolution to a chain of small-scale, step-by-step processes. These step-by-step processes on a small scale have even been replicated in the laboratory, for example with the famous fruit fly *Drosophila*, of which 25 generations can be cultivated in one year. This way, species could be seen to change significantly within a short period of time by selecting organisms for certain characteristics.

Imagine, artificial evolution right before your very eyes! This process is called micro-evolution. It is the basic process of evolution for population geneticists and evolutionary biologists. How do we imagine this process to work, and what role does chance play in it?

Selection only works if there are differences between organisms, otherwise there is nothing to select. Moreover, these differences must be hereditary if the effect is to be visible in subsequent generations (for example, selecting horses with clipped tails does not produce foals with clipped tails). Well, such hereditary differences arise from mutations, and those mutations arise by chance. What does that mean?

Mutations do not arise on request but "just happen," regardless of what is needed or beneficial. So they are accidental in the sense that they occur in an *unintended* or random way—as when someone blindly draws lots and then must wait to see what the outcome is. They are also coincidental in the sense that it is unknown and *unpredictable* which mutation will occur and when. We do know that radiation and chemicals can increase the chance of mutations, but we cannot predict which mutations will in fact occur. Mutations are thus not only unintended but also unpredictable. In that sense, they are coincidental in two different ways.

Whether they are also accidental in the sense of being *undetermined* is still the question. Some biologists, such as Jacques Monod, argue that a mutation is a quantum event, so it would be subject to the Heisenberg uncertainty principle. In that case, a mutation is unpredictable. But it is of course also possible to say that a mutation is in fact determined, but that we still know too little about the lawful mechanism behind it. For the time being, it is not clear which position is the correct one. That is why we must be careful about the coincidence of mutations: they are unintended, unpredictable for the time being, and perhaps undetermined. Those are three different "types" of coincidence.

What happens after these mutations have occurred? Do they end up in the next generation(s)? That also depends on a number of chance factors. In organisms such as humans who possess all hereditary factors in duplicate, there is only a 50% chance that the mutated factor will end up in the sperm or egg cell. Whether the cell with the mutated factor will subsequently fuse with another cell is again a matter of chance. In such cases we again assume some kind of lottery—it is, as it were, playing dice for posterity. The "randomness" lies in a blind, random, or unintended draw, where there is no preference for a certain outcome in the game. But that does not mean that these processes are also coincidental in the sense of being unpredictable (because the probabilities can be expressed in percentages).

## *Preferred coincidences*

But with that, we have discussed the main types of coincidences in evolution. Does that mean evolution is entirely ruled by coincidences? Not really. All organisms (with all their hereditary factors) must pass through the sieve of natural selection—and that "sieve" does not work randomly. Some organisms, compared to their peers, have relatively better chances of surviving and reproducing, passing on their hereditary factors to the next generation proportionately. Natural selection therefore has a preference for certain organisms and is therefore by no means an arbitrary process. And yet many biologists call natural selection "accidental." But in doing so, they give a new meaning to the concept of chance. What they mean is that natural selection is a short-sighted and *opportunistic* selection process. Natural selection is based on current environmental requirements—fast and efficient, but also potentially short-sighted. Whether a particular adaptation is also beneficial in the longer term is irrelevant to natural selection.

Where do all these coincidences lead to in evolution? In microevolution, they lead to changes in the genetic composition of a group of organisms; this has even been shown in the laboratory. But evolution requires more. And with that we leave the microdomain of the population biologist and enter the realm of the evolutionary biologist. The latter tries to explain how new species do arise in evolution. They assume that species can be separated from each other by what they call *reproductive* barriers. How can we now take the step from microevolution to speciation? In other words: can two populations of one species change their inherited composition in such a way that two species arise out of one?

We've never seen anything like this happen in the lab. Yet most evolutionary biologists assume that microevolution can eventually lead to speciation. They often assume that a *geographical* barrier has occurred between two populations by chance, so that without further contact with each other these two populations grow apart so much that after a while they no longer "recognize" each other when contact is renewed. As a result, they can no longer mate. The geographic barrier has then become a *reproductive* barrier. This is where a new chance factor has come into play: the chance occurrence of a geographical barrier.

Apparently, evolution—at least micro-evolution and speciation—does depend on coincidences. Does that tell us everything about evolution? There are still biologists who claim it does not. These are mainly the paleontologists among them, who view evolution not on a small scale but on a large scale. They see evolution primarily as a macro-evolution: not only as a gradual change of genetic composition and not even just as a division of species, but also as a large-scale transformation process, in which drastic structural changes and transitions can occur—for example, from reptile to mammal. Is macro-evolution really nothing but a succession of micro-evolutions—step by step, gene by gene? And if so, how can the many coincidences of a microevolutionary process produce something as impressive as the rich family tree of life?

### Evolution on a large scale

The problem is that the accidental—that is, unintended—nature of mutations and the accidental—that is, opportunistic—nature of natural selection, known from microevolution, are almost impossible to explain macroevolution. Instead, we see that macroevolution is moving in one direction. If we were to play a movie about the evolution of life backwards, it would make a very strange impression. We would immediately notice that evolution is going in the "wrong" direction. But would we instead film the micro-evolution of those well-known white moths on white birch trunks, which are both light-colored at first, then selectively darken due to air pollution, and then lighten up again when the polluting

## Chapter 6: The Evolution of Life

industry is pushed back, we could very well run this movie backwards without anyone noticing. Micro evolution can go in either direction.

Macroevolution, on the other hand, seems to be moving in one particular direction. What then is the "direction" that we see in the "film" of life on earth? Not only are we seeing an increase in species (although there are also many species that have become extinct), but also an increase in complexity—that is, more variety of body parts and more variety of bodily functions. Complexity is, as it were, a criterion for measuring the progress of evolution. Incidentally, there are also types or species of organisms that have hardly changed at all in millions of years, the so-called living fossils, but most branches of the family tree of life have undergone profound changes, accompanied by increasing complexity.

If there is indeed such a thing as an increase in complexity, then the next question is what the driving "force" is behind this process. Is it a "propulsive force" or rather a "suction force"? To put it metaphorically: is the direction of a river's flow determined by the attractive force of the sea or rather by a propulsive force along the path of least resistance?

It is very tempting to assume an attraction force in macro evolution, which would draw the evolutionary process towards a certain goal. That would mean not only a direction but also a goal. Evolution is therefore often regarded as a development from "lower" to "higher" forms of life. But it's so hard to pinpoint what the "endpoint" is. Some see the perfection of life as the ultimate goal, but that raises the question of what the perfect life should look like?

Others see a clear growth towards more individuality (Julian Huxley) or even towards more (self-)awareness. Others see the ultimate goal as the culmination point called Omega (Teilhard de Chardin). But is that really a form of attraction that is active and visible in macroevolution? Is this still biology or should we speak of a philosophy, or philosophy of life, which is then projected into the evolutionary process? Moreover, such concepts are too vague and general to be testable.

That's why it seems to me we should be content with less—not an attractive force, but a propelling power. As water flows in the direction of least resistance, so life flows in the direction of better survival, better reproduction, and in the long run more complexity, efficiency, flexibility, independence, more control over the environment, more awareness of the environment. A more general description would go like this: evolution is a process towards a better organized and therefore better functioning system. According to this view, large-scale evolution is like a large stream that "meanders" based on small-scale evolution. This way, it eventually ends up "at the sea."

The big question now is how micro-evolution, with its many coincidences, can steer the great stream of macro-evolution in the right direction. How can so much coincidental disorder create so much order? How is it possible for coincidences in micro-evolution to lead to macro-evolution?

## Chapter 6: The Evolution of Life

### *The script of evolution*

We have already seen that natural selection is accidental in the sense of being opportunistic, but not necessarily in the sense of being unintended, unpredictable, or undetermined. Accidental mutations create a supply of abundant and hidden genetic potential that future generations can draw from under unforeseen circumstances. These mutations thus provide the raw material for evolution, from which natural selection can selectively draw as an ordering and directing factor. This is how order may emerge from disorder. Moreover, this selection process is cumulative—that is, it is a process with a "memory": whatever is sifted out has already been sifted out in previous processes.

Does this mean that the course of evolution had to go the way it does? We don't really know, certainly not in strict scientific terms. Perhaps the best we can say is that the evolutionary road to more complex forms of life is a process that meanders like a river. On the one hand, it "meanders" by following a path that seems coincidental and random. On the other, in spite of its winding flow, it also moves in a specific direction, steered by natural selection, which is like a path of least resistance. Just as a river follows a path of least resistance according to the topographic design of the landscape, so does the "stream" of evolution follow a path somehow regulated by the design of our universe and its organisms. Accidental but far-reaching disturbances of this subtle equilibrium have little chance of success. Anything that deviates too much from that direction is nipped in the bud. This is why certain types of organisms develop in a certain direction.

However, the picture of evolution as a meandering river is, on closer inspection, not such a good idea. Organisms are not as malleable as water, for they have their own "structure." And with that, the contours of the evolution scenario have been indicated. The unintended chance process is streamlined and channeled both by the "bed of natural selection" and by the "structure of the life forms." With a different image: chance follows the bed of necessity. Natural selection favors what is successful but hinders what is not. To be successful, organisms must follow the laws of nature. For instance, fish could not swim well if they didn't follow hydrodynamic laws, and birds could not fly well if they didn't follow aerodynamic laws. The spontaneity of chance is therefore restricted in evolution because of certain constraints found inside the organism and inside its environment.

### One last pitfall

In the second part of this chapter, we talked about the concept of coincidence or randomness as it is used in science. This "coincidence" is mainly unintended, often opportunistic, sometimes unpredictable, but rarely undetermined. Outside science, however, "coincidence" has a very different meaning: it is the designation for something like a mysterious force that works outside of natural causes. That is a form of coincidence preached by the worldview of fatalism. What is called "accidental" in this sense is not only unpredictable and unintended, but also meaningless.

Behind the conception of fatalism is a philosophy of life that has nothing to do with the life sciences. The natural sciences simply do not deal with questions of meaning and purpose. Nevertheless, this does not prevent some biologists from pursuing a fatalistic view of life. However, when they do so, they have given the concept of "coincidence" a completely different, unscientific meaning. The question of whether evolution also has "meaning" or "purpose" is not a scientific question, but one of a philosophical and/or theological nature. It is impossible for natural science to study that aspect—although some biologists would still like to do just that.

Scientists have no right in their role as scientists to decree that we as human beings are unintended, unplanned, unguided, fortuitous creatures, or mere products of a blind, meaningless, and purposeless fate. Science has nothing to say about such issues. Evolutionary theory cannot even explain why there is evolution to begin with. Can we still talk about evolution in terms of purposes and the like? Could evolution still be seen as steered by the purposes the Creator has in mind? I don't see why not. The case could even be made that evolution cannot be understood without any reference to its Creator, God.

If we do assume or declare that the theory of evolution completely explains the course of evolution from simple organisms to complex human beings, then we get into serious trouble, for now the question arises as to where that theory itself comes from. The answer many will favor is that the theory of evolution originated in the mind of Charles Darwin, and now still resides in the minds of many biologists. But then we must question how trustworthy this theory is, given the fact that this very theory must then also be a product of evolution. Doesn't that outcome defeat its own claims? Doesn't that lead to self-destruction? If we claim that evolution is all there is to life, then we undermine all our own claims about evolution. If we claim that all we know about this world is merely the outcome of evolution, that is the downfall of all our claims. Our claims *must* be more than evolutionary products if we want them to be true and valid.

# Chapter 7

## Is Man Stripped of his Clothes?

Not too long ago, I heard on a TV program that science has "stripped man." It would be "thanks" to science that man has been robbed of his imaginary "clothes"—that is, his worth and dignity. The central question was: what is left of man since science came along? Many would answer something like this: nothing but a glorified ape, nothing but a computer of flesh and blood, nothing but a simple piece of DNA, nothing but a mishmash of molecules! If people today are easily treated like garbage, then that is mainly the fault of science, so the reasoning goes.

The situation reminds me of the tale best known as "The emperor has no clothes." Two swindlers pose as weavers, who offer to supply the emperor with magnificent clothes that are invisible to those who are stupid or incompetent. After these two weavers report that the emperor's suit is finished, they mime dressing him and putting him in a procession before the whole city. No one wants to appear inept or stupid, until a child blurts out that the emperor is naked, wearing nothing at all. The people then realize that everyone has been fooled. That's how we should feel after science has supposedly shown us the truth of our nakedness. We are stripped and ripped off.

### Knocked off Man's throne?

When did that happen? It all started in the early stages of science—to be more specific: it began with Galileo Galilei. By him, Man was silently dethroned from his throne. Since Galileo, Man, as an inhabitant of Planet Earth, is no longer at the center of the universe. And then the onslaught of science went even further. Another blow to Man's dignity was dealt by Charles Darwin: thanks to the theory of evolution, Man lost his special place in creation. And with that, Man also lost his monopoly position on earth. And if that wasn't enough, Watson and Crick came up with the message that the secret of Man can be traced back to a long chain of DNA molecules. These and other discoveries seem like parts of a global conspiracy of science against humanity, changing Man into a mere man or woman. At least... that's the view of some people. But before we get carried away in this train of thought, we should first consider to what extent science is rightly being blamed. In other words, what has science done to Man?

### Science undresses

Anyone who has ever done some research knows that in such a case you can never pay attention to everything all at once. First of all, you focus your interest on certain aspects of the world around you. A psychologist looks at things differently than a biologist, and a biologist sees different things than a physicist. In a sense, each discipline therefore creates its own "world" of things and facts. As a result, a psychologist has an eye for psychological processes and matters, while a biologist observes biological facts and events. This is not to say that our world is compartmentalized, but any given process can have many aspects, not just psychological or biological, but many others as well. The same event can therefore be viewed through different "glasses" or with different "spectacles."

That's a first step. Then, in each experiment, attention is focused on what is considered "relevant." For example, during the experiment that you are conducting, you may specifically focus on a change in color, temperature, direction, or movement. In other words: there must be something like a problem statement. This accurately defines the researcher's field of interest. Great scholars can be recognized by their ability to reduce their many question marks to a more manageable problem.

Each scientific research therefore requires demarcation and limitation. Good researchers are masters in defining their problem. But that's not all. In addition, they must ensure that the event to be investigated is "lifted" from its natural context. Technically speaking, the experiment should eliminate interfering variables and keep all other variables under control. Experiments take place, as it were, in the protective test tube of the laboratory, so that nature cannot throw a spanner in the works. A biological experiment, for example, is simply more difficult to control in the body (*in vivo*) than in a test tube (*in vitro*). In short, a problem must not only be defined, but it must also be made manageable and controllable.

The consequence of all this is that science works with and assumes an artificial "world"—that is, a world with selected things, events, and phenomena, which are also manipulated in some way. It is often said that science by definition works in a *reductionist* way: the complexity and multiformity of events is "reduced" to a manageable model with an analyzable problem. Typically, science strips down everything it investigates.

## Mapping the world

Quite a lot follows from this for a correct assessment of the results of scientific research. In one way or another, scientists are "map makers" who map our world with their special techniques. Sometimes such a map can be very useful, provided you have the right map. After all, a railway map is useless when you are driving a car. It is important to realize that each science produces its own kind of maps—something for everyone: a geographical, a physical, a chemical, a biological, a psychological, or a sociological map.

Against this background, it becomes pointless to ask which map is the best map. It just depends on what you want to do with the map. Each map states what is relevant within that particular setting, thus omitting what is considered irrelevant. What is missing on such a map is therefore not denied, but simply omitted or neglected. That is why a map can never replace a landscape. A map that shows all the details of the landscape would be as complex as the landscape itself—and then it is no longer a map. Nevertheless, a map remains useful if we are looking for certain details during our journey through the landscape.

# Chapter 7: Is Man Stripped of his Clothes?

It is only in this light that we can understand what scientists have done with Man and his world: they have mapped everything that is relevant to a particular purpose as closely as possible—and that work is still far from finished. Galileo was the mapmaker who designed an astronomical map of our world. Less notorious designers would follow him. Next, Darwin came up with a biological map of living nature. In the past, man had to learn by experience to find his way; now he has been handed various kinds of maps on many different scales to guide us on the way. And as such, these maps can be helpful to us, because they open up a new world for those who can read them—or rather, a new view of the old world.

**Man is nothing but...**

It is important to discover that science works with maps and models which are abstract representations of reality, eliminating what does not fit in that particular approach and at that particular scale. Whatever is outside the model is only "provisionally omitted"—but that is quite different from "denied for good." We are therefore no longer surprised that there are no paving stones or human beings on astronomical maps. They are not denied but simply omitted! Anyone who says that they "therefore" also do not exist is going beyond his or her book. Such a person forgets that a "landscape" is something more than any particular "map."

This has quite some consequences. To study humans, for example, a DNA map can be made. But that is quite different from saying that Man is the same as his DNA map. And yet that is really the idea behind the slogan that man is nothing but his DNA—or that man is nothing but stardust (as Carl Sagan put it), or that man is nothing but a glorified ape (as Charles Darwin put it), or that man is nothing but a pack of neurons (as Francis Crick put it), or that man is nothing but a bundle of instincts (as Sigmund Freud put it). Seen that way, it seems as if only one map could be made of Man, with the denial of all other aspects, angles, and perspectives.

Whatever applies to the map, therefore, does not cover the entire landscape. What is acceptable within the framework of a certain model or what is acceptable within a certain approach cannot simply be applied outside these frameworks. Anyone who claims that everything (!) is the result of natural selection à la Darwin, or that everything is the result of libido à la Freud, or of socio-economic forces à la Marx, must realize that then also the theories of Darwin, Marx, and Freud cannot be an exception to this. And with that, they have undercut themselves! The requirement of reduction is a sound and necessary rule for map making; but the landscape itself is beyond this reduction.

### Which destination?

Once we understand the difference between the "landscape" and its "maps," the problem is not so much that we do make maps of the landscape, but it is important to ask ourselves how those maps are going to be used. Opposition to science is growing on many fronts, among other things because the findings of science are said to be an attack on human dignity. For example, there is

# Chapter 7: Is Man Stripped of his Clothes?

still criticism of the neo-Darwinian map from a religious perspective; and there is increasing opposition to the neurological maps of human behavior from a psychological angle. However, I think it's not right to target the maps, but instead the attack should be directed at the developers behind the map makers.

Let's take an example. Aggression in human society is a phenomenon that has been mapped in many ways. According to a common biological version (from biologists such as Konrad Lorenz and Desmond Morris), aggression is programmed in "the nature of the beast"—according to a kind of three-stage rocket: that's how our brains work, because that's how it's stored in our genes, and it's those genes that we owe to the evolutionary process our ancestors went through. Does this completely explain human aggression? Probably not!

Why not? Because there is also a sociological "map" of aggression. According to this view, aggressive and violent behavior of people can be explained because of environmental influences. Aggressive behavior can be learned in an aggressive environment, by imitating aggressive model figures from the immediate environment or from the TV screen. Does this completely explain human aggression? Probably not!

Why not? Because there is also a psychological version. Someone who uses this map is more likely to look for the cause of aggression in terms of personal frustrations. It is those who are frustrated in satisfying their needs or in achieving a desired goal will respond with aggression. Does this completely explain human aggression? Probably not!

The end result is now: at least three different stories that chart the phenomenon of aggression in three different ways. Of course, to resolve their differences, we should acknowledge that the biological version tells us how such a thing as aggression arose, the sociological version explains why one group of people is more aggressive than another, and finally the psychological version explains us what makes one person more aggressive than the other. But this solution leaves us with a practical problem. Those who want to combat aggression will almost inevitably base their therapy on the mapped diagnosis. When it comes to concrete applications, the use of only one map can be dangerous because it ignores many other factors that are not covered by that one particular map.

Yet a map as such is harmless: it tells us *how* we could go to get somewhere else, but it doesn't tell us *where* to go. That is its strength and its weakness. After all, a map never contains its own destination; the destination is in the minds of the creators and users of a map. It will always depend on us, on what we will do with our knowledge of the "stripped" man. Hence, the evil is not in science itself, but in what is done with the help of science. Science did not strip Man of his clothes. That was done by people who would like to make that their personal destination.

# Chapter 8

## Ape or Adam?

Human beings are animals like all other animals: they all breed, feed, bleed, and excrete. Even St. Thomas Aquinas did not need any learned biology to see the plain truth that humans are first of all animals. Obviously, Aquinas did not know about evolution and evolutionary biology—those concepts would be conceived some six centuries later—yet, he could see humans as part of the animal world. But at the same time, he distinguished them as unusual and very special animals who have the faculties of reason and intellect. Therefore, he called a human being not just an animal but an *animal rationale*.

Obviously, distinguishing humans as special animals does not imply that they came forth from the animal world. The same can be said about classifying humans as Primates or Apes. They may have many features in common, but that does not mean they came forth from other Primates or Apes. Yet, that is what most biologists nowadays would claim.

According to the theory of evolution, the ancestors of man are not to be found in Adam and Eve, but among Apes. That is the message biology seems to tell us after years of research. The living world is characterized by an enormous diversity—from plants to animals, from insects to mammals, from mice to elephants—and yet there is a striking unity: They all share many communalities and seem to be part of one gigantic family tree. Accepting this idea would make much more sense of the way we ourselves fit into a classification of the animal world.

If this is true, then humans too seem to have many "relatives" in the animal world, with some looking like close relatives and others like more distant relatives. In fact, humans have a skeleton like some of the animals, the vertebrates, do. In addition, they start their lives in a womb, just like some of the vertebrates, the mammals, do. Then they have a relatively large brain, which they have in common with a subgroup of the mammals, the primates. And like some of these Primates, the Hominids, they lack a tail. When tightening the range, the resemblances become more and more numerous and striking. This makes it easier for us to see that we all are part of the same family tree that connects all that lives on Planet Earth.

Nowadays, there is hardly any biologist left who doubts the correctness of this message. Why not? Because the theory of evolution gives us a plausible and credible biological explanation of the wide variety of life forms in nature, including Man. Its statement is basically simple. For most traits, hereditary variants (mutations) either exist already or emerge; these variants differ in viability and are thus passed on to the next generation to a greater or lesser extent, which is called natural selection. As a result, the hereditary composition of a population changes, so new species arise, and complex life forms can arise from simpler forms. That's pretty much the biological story based on the mechanism of natural selection.

There are many scientific indications that this selection mechanism does indeed work in nature. Current biology is practically unthinkable without the assumption that there is such a thing as evolution and that it works on the basis of natural selection. As the late geneticist and evolutionary biologist Theodosius Dobzhansky famously said, "Nothing in biology makes sense except in the light of evolution." However, it is up to discussion whether natural selection is the only mechanism of evolution, but that there is evolution—that is, that all organisms have common ancestors—is something hardly any biologist today will dispute.

Is evolution a proven fact? Not completely. There are still many questions unanswered, as we discussed already. For some, this has led to a complete rejection of evolutionary theory. One group of them is made up of creationists—they don't believe in evolution. They have therefore developed an alternative theory that is not based on the theory of evolution but on creation—the so-called "creation theory." Although this theory uses a number of scientific data from paleontology, it is mainly based on religious statements from the Bible. Incidentally, it is striking that creationists prefer to quote chapter 1 from the book of Genesis, and hardly ever cite chapter 2, in which an alternative creation account is presented. This led them to what is called *creationism*.

## Creationism

Broadly speaking (because there are many, many variants), the creation theory boils down to this. "In the beginning"—a little over

10,000 years ago—God created some basic types of organisms that lived in harmony with each other and with their environment. Therefore, no fossils were formed at that time. But afterwards two major events occurred: the Fall and the Flood. As a result of the Fall, God changed the behavior of certain animals, and only then did predators arise and did death enter the scene.

Because all types of organisms were created in a very short time, different biotic communities must have existed side by side, for example a Devonian landscape next to a Cretaceous Sea. Moreover, due to the "greenhouse effect" of a protective and invigorating atmosphere, the organisms were larger and more numerous than they are today. The fact that we have found remains of these organisms is a result of the great world catastrophe of the Flood (dated 5000 years ago). Remnants of that old world were stored in layers of the earth at the time and these fossils are now being excavated by paleontologists. Incidentally, there were also mammals then, but they were able to flee the rising waters of the Flood for a while—that's why they can only be found in higher layers of the earth. So much for the core of creation theory.

Why do creationists have so much trouble with the phenomenon of evolution? Why do they insist that there is no evolution? Could it be because the Bible says nothing about evolution? But that is not a strong argument, because the Bible says nothing either about the orbits of the planets and many other natural scientific phenomena. Moreover, as we said already, the Bible has two different creation accounts. Is it then perhaps because evolution is, by definition, contrary to creation? But you don't have to accept that false contradiction. At the time of Darwin, some theologians were already saying that evolution might be God's way of creating.

Even the claim that the world was changed by the Fall of Man is rather questionable. Were there "thistles and thorns" before the Fall? And were animals different before the Fall took place? Probably not. St. Thomas Aquinas made a very astute remark in this context centuries ago: "some say that the animals, which are wild now and kill other animals, were not that way [in paradise].... But this is entirely unreasonable. The nature of animals was not changed by the sin of man."

*Evolutionism*

I think there's a more important reason for rejecting the theory of evolution. When some biologists (and many others with them) talk about evolution, they're not talking so much about the multiformity of life and its biological explanation, but they often refer to something much more extensive. Which is a worldview, best called *evolutionism*.

This worldview uses the theory of evolution to explain not only biological phenomena, but also all sorts of other phenomena that fall outside the biological scope, for example data from the fields of sociology, psychology, ethics, or religion. Behind this view is a deep conviction that not only the multiformity of life, but also our society, our morals, and our beliefs are the result of natural selection. In the meantime, these extremists forget that this claim is a form of suicide, because their belief must then also have arisen from natural selection. And that is a very shaky and unreliable basis for such a far-reaching statement. This is an example of what G.K. Chesterton called "the suicide of thought." If science is all that matters, then science becomes a pretty shaky enterprise, resting on man-made quicksand. Scientists would fundamentally lose their reason for trusting their own scientific reasoning.

Because these extreme kinds of claims about evolution have very far-reaching pretensions, which far exceed the competence of biology, the term *evolutionism* is quite appropriate. The ending -ism is in fact a common designation for positions that are philosophical, metaphysical, or ideological in nature. Most "isms"—such as materialism, scientism, relativism, secularism, and the list does not end here—are more than harmless techniques; they arise from what are in essence ideologies, doctrines, worldviews, or belief systems. They look at the world from one specific perspective, pretending this makes you see everything and all there is. Well, here we have such an ideological point of view. Evolutionism uses the theory of evolution as a theory of anything and everything in nature; it promotes an evolutionary explanation as an all-encompassing explanation.

Because evolutionism exploits the theory of evolution as an all-encompassing explanation of this world, it is a form of biologism that has overstepped the boundaries of its own methodology and has therefore assumed the status of a philosophy of life, a worldview, an ideology. A biological theory about the origin and evolution of life is blown up into a philosophical view of life about anything and everything in the animate world, pretending that biology is the mother of all sciences.

That is not a new development, by the way. At the end of the last century, the philosopher Herbert Spencer had already realized that Darwin's theory of evolution lends itself perfectly to a new view of life and the world. Darwin's principle is so simple that Spencer thought he could apply it to just about anything to create an orderly world. That is why he wrote a series of books in which he used the principle of natural selection to explain psychology (*The Principles of Psychology*), sociology (*The Principles of Sociology*), and ethics (*The Principles of Ethics*). He thus turned biological Darwinism into social Darwinism. The latter implies that human society is nothing more than a battleground in which the adapted survive at the expense of the non-adapted. Before Nietzsche came into play, the idea of a superhuman race was already born.

## *Formidable opponents*

Not surprisingly, as soon as Darwinism is elevated to a philosophy of life, it meets creationists on the battlefield. Because creationism is also a philosophy of life, evolutionism and creationism are both completely opposite worldviews. What they have in common is that they both believe they can provide a comprehensive explanation for everything in this world, including the sociological, psychological, and ethical aspects of human life—and therefore in both cases, we must speak of a philosophy of life.

The battle between evolutionists and creationists has been going on for some time, especially in the United States. They criticize the scientific merits of each other's theories. The question, however, is whether the background of this contradiction should not be sought at a deeper level. What they actually criticize is each other's pervasive philosophy of life. They oppose the monopolistic claims of their opponents, but at the same time they are guilty of the same offense by overlooking their own misconceptions. What then are the shortcomings of both worldviews?

One might say the following to evolutionists. Whatever the theory of evolution may teach us about the actual mechanism of evolution, such a theory at best provides us with a biological story, a biological view of things. Biology, like the other natural sciences, searches for causes and functions. But there are other searches. Religion and theology, for instance, search from a different angle. They do not look for causes, but for the grounds of our existence and of the world as such. And that difference is quite fundamental. God is the ground of my existence, but the cause of my origin lies with my parents.

Apparently, in addition to the angle of biology, there is the angle of theology and religion. The angle of biology portrays the evolution of life. People who look for the *causes* behind the multiformity of life will probably end up with the evolutionary theory of natural selection. But seen from the angle of theology and religion we will probably soon come across the creation account of the book of Genesis. What is central in that account is not *how* man came into existence, but *why* man came into existence. It is not about causes but about grounds and foundations. Unfortunately, evolutionism denies this last type of question. It thinks it has an answer for everything and denies that there are questions to which the theory of evolution has no answer. Evolutionism does not willing to acknowledge how limited the theory of evolution is.

But things are not much better for creationism. Something similar can be said about this worldview. The doctrine of creation deals with our existence and destiny—which is different from our origin and descent. Creationism basically denies that questions about *how* man came into existence are different from *why* man came into existence. Creationism therefore has little to do with scientific insights into causes and effects. St. Augustine already realized that creation is not a causal affair that is taking place in time. As he put it, "the world was not made in time, but simultaneously with time." The consequence of this is that "to create" does not mean "to cause" but rather something like "to bring into being and existence." Creation is different from evolution, because evolution presupposes creation. Searching for the origin of life implies that life can exist. Creation is not evolution, and evolution is not creation, for creation is the condition of evolution.

Therefore, there is no reason to choose between creation and evolution, and therefore no reason to choose between creationism and evolutionism. They are both extreme worldviews. They are both more concerned with their own narrative than the truth. But if that is true, there is also no reason to choose between ape and Adam. It's not either-or but both-and. When it comes to the causes of our origin and descent, it is probably better to look for our ancestors among the Apes. But when it comes to the grounds of our existence and destiny, it is best to turn to Adam and Eve.

**What about Adam and Eve?**

What creationism basically ignores is the second creation account in which Adam and Eve feature. These two have something to tell is that evolutionism plainly ignores.

Whereas the first creation account tells us how beautiful God's creation is, the second account explains why there is also something very wrong with what God had created so beautifully. But that latter part is not God's doing—it is ours, or in fact Satan's. Adam and Eve are a pivotal part of creation. First of all, they unite all of humanity. All human beings and even all human races are united with each other through Adam and Eve. These are our common ancestors. The human race forms a unity that transcend all different races. This creates unity in the midst of diversity.

Second, Adam and Eve are also pivotal to explain what went wrong in creation. What we also have in common with all of humanity is being part of the ancestry of Adam and Eve who "spoiled" everything for us. How did they "spoil" it for us? C.S. Lewis put it well when he wrote, "We are not merely imperfect creatures who must be improved; we are rebels who must lay down our arms." Adam and Eve became rebels and we follow their example. It's called the Original Sin that we "inherited" from them, similar to the way dysfunctional families create new dysfunctional families.

As a matter of fact, the human race forms a unity—not only in its origins, but also in its sinful state. The descendants of Adam and Eve were distorted by sin because their first ancestors chose to distort themselves. Although the first humans, Adam and Eve, were given the faculties of rationality and morality, they deliberately chose to follow their old animal drives. Blaise Pascal said it right, "It is dangerous to show man in how many respects he resembles the lower animals, without pointing out his grandeur." But Pascal emphatically added next, "It is also dangerous to direct his attention to his grandeur, without keeping him aware of his degradation"—which is a reference to the Fall.

Third, the doctrine of the Original Sin is essential to Catholic Faith for another reason. If there is no Original Sin, then the Cross is a hoax; and if there is no Cross, then the whole economy of Salvation through the Incarnation is up for grabs. The Catechism puts it this way, "The Church, which has the mind of Christ, knows very well that we cannot tamper with the revelation of original sin without undermining the mystery of Christ." Clearly, if humanity is not in fleshy solidarity with itself due to a descent from two common ancestors, then Christ's fleshy solidarity with all of humanity is rendered problematic.

No wonder, Adam and Eve keep standing. Pope Benedict XVI summarized all of this during a conference at Castel Gandolfo in 2008,

> *The clay became man at the moment in which a being for the first time was capable of forming, however dimly, the*

*thought of 'God'. The first Thou that—however stammeringly—was said by human lips to God marks the moment in which the spirit arose in the world. Here the Rubicon of anthropogenesis was crossed. For it is not the use of weapons or fire, not new methods of cruelty or of useful activity, that constitute man, but rather his ability to be immediately in relation to God.*

# Chapter 9

## Is Everything in the Genes?

The study of genetics has been quite successful in tracing *many* features of our body back to genes and their DNA. If we keep searching, will we someday be able to find genes for *every* feature in the human body? It is certainly a fascinating question. Imagine, every feature we have—not only features such as skin color but also features such as free will, rationality, morality, and religion—will eventually be explained by genes.

It is certainly something most geneticists, if not all, are striving for. To name just one of them, Sidney Brenner, a pioneer in DNA research, said not too long ago he could compute an entire organism, humans included, if he were given its entire DNA sequence and a large enough computer. With a like mind, the American molecular biologist Walter Gilbert had the audacity to claim that "when we have the complete sequence of the human genome, we will know what it is to be human." And James Watson, the co-discoverer of DNA, trumpeted with a similar audacity to everyone: "We used to think that our fate was in our stars. Now we know, in large part, that our fate is in our genes." At least he was honest enough to add the clause "in large part." Yet it should be stated that what these geneticists said about genetics goes far beyond genetics and their expertise as geneticists.

Let me just make a logical remark first. Claiming that *all* human features can be explained by genetics—as has supposedly been done already for many of them—is not a logically valid way of reasoning. It is called generalizing induction, which is supposed to take us from a general statement about "many" cases to a universal statement about "all" instances. But this is not a logically safe or validated conclusion. It is at best a belief or conviction. Inductivism was ridiculed by Bertrand Russell with his story about the "inductivist turkey," that had been fed on the farm for many months at 9:00 AM. That made the turkey rely on this repeated pattern until Christmas day came along!

At least, logic can help us recognize that we are dealing here with a claim of a dubious kind. Inductive reasoning cannot even be justified by citing that induction has already been successful in many other cases, for that would make for another inductive argument. However, this logical objection does not get us far in our discussion here. Showing that certain human features have not been connected to genes yet does not rule out they might be someday in the future. The fact that a certain feature has no known genetic roots does not entitle us to claim that it *cannot* have genetic roots. To prove the latter claim we need stronger arguments.

### An a priori way

Sometimes the tools of logic and philosophy can tell us ahead of time—*a priori*, if you will—what science can achieve and what it

## Chapter 9: Is Everything in the Genes?

cannot. Let me explain this with the following case that I borrowed from the Nobel Laureate and physiologist Peter Medawar.

Anyone looking at paintings of the famous Renaissance painter El Greco will notice that most of his figures are unnaturally tall and thin. Some scientists were eager to explain this in scientific terms. One of them came up with the hypothesis that El Greco must have had a form of astigmatism that distorted his vision and led to elongated images forming on his retina. Sounds interesting, but this hypothesis is doomed to fail from the very beginning. Even if El Greco did see the world through a distorting lens, the same distortion would apply to what he saw on his canvas. These two distortions would cancel each other out, and the proportions in his pictures would remain accurate. So, we must come to the conclusion that El Greco's figures, particularly the holy ones, appear unnaturally thin and tall because that was his intention, not forced by any abnormality or gene—he had painted them that way on purpose.

Imagine that these scientists had spent lots of time, energy, and funding on testing their hypothesis, not realizing that it was bound to fail ahead of time on purely philosophical, or at least logical, grounds. Yet, some did not give up that easily and tried to attribute Vincent van Gogh's preference for yellow colors to a visual disorder called xanthopsia; others mentioned drug use or glaucoma. Even if Van Gogh did view the world through a yellow filter, he would also view the colors on his canvas that way. Apparently, he, too, had chosen the yellow color on purpose!

In all such cases, science just fails us, since philosophy and logic are able to show us *a priori* that the above scientific explanations are doomed to fail, even before any scientific test has been done. There are certain "scientific" ideas you should not spend your time and energy on, because philosophy and logic can show you that they are headed for a dead end. They are at best "inventions" that could not possibly make it to "discoveries," purely on logical or philosophical grounds.

All scientific endeavors to anchor certain human features in genes or DNA might also be doomed to fail ahead of time for mere philosophical or logical reasons, as I am going to show you. Sometimes we know ahead of time—without doing any genetic research—that, in cases like these, a genetic explanation does not make sense or is not even possible. Let me demonstrate this for what is usually called the "free will" of human beings.

Could our free will perhaps be based on a gene or a series of genes? Theoretically, there could indeed be a gene that allows us to make choices, but if this gene, or additional genes, would also determine the outcome of these choices, then we cannot really make free choices and have basically lost the free will we thought we had. Is it possible to eliminate our free will through genetics? If we could, we would end up with what is called *genetic determinism.*

Philosophy and logic may help us to answer this problem ahead of time, "a priori," without us having to do any research in genetics. One of the strongest arguments is that the world view of universal determinism makes for a contradictory claim. If genes, for instance, really determine everything in my life, then they would also determine my choice to believe, or not to believe, that genes determine everything. The key problem is that we are dealing here

## Chapter 9: Is Everything in the Genes?

with beliefs, and beliefs are not material entities like genes—unlike genes, they can be true or false. If I believe that genes determine everything, I have no reason to suppose my belief is true, and hence I have no reason for supposing that genes determine everything. Determinism is a "boomerang theory" in optima forma—it defeats itself, for once we consider it to be true, it becomes false.

Apparently, we are dealing in this discussion with *thoughts*. Unlike neurons and genes, thoughts are *immaterial* entities. If our thoughts—including those of scientists—were merely neural activities and genetic products, they would be as fragile as the molecules they supposedly came from. That would surely be the end of science—and of all the genetic issues we are talking about in this chapter. If the science of genes were the mere product of genes, it would inevitably be a self-defeating enterprise. So let us avoid that cliff!

In other words, there must be something more than genes and neurons to regulate our behavior: our thoughts before anything else. Whereas impulses come from the brain and from all that comes with brains—genes, hormones, and the like—thoughts come from the mind. When humans deceive someone, for instance, we tend to look for motives in the minds of suspects, not for defects in their brains, let alone in their genes. As human beings, we live in two worlds: we are driven by physical causes (hormones, reflexes, drives, genes) as well as by mental causes (motives, reasons, intentions, thoughts, and beliefs). But these two categories harbor very different entities: mental causes are different from physical causes and are not under their control.

So, our conclusion must be that the free will of human beings is connected with the world of thoughts, rather than the world of genes. That's why a genetic explanation of the human free will is logically unacceptable. And something similar can be said about the human features of rationality, morality, and religion. That's why scientific or genetic explanations are doomed to fail on *a priori* grounds.

Perhaps I should mention one caveat. All free choices have to work within existing boundary conditions—since we cannot effectively choose the impossible. So, our freedom of self-determination does not let us do whatever we want to do. The more we are aware of our constraints, the more we can actually be free. "Know thyself" were the words written above the temple doors of the oracle at Delphi. With the proper knowledge, we can take charge of our constraints so that we are no longer their victims, but rather their architects. That is why I said earlier that we can, to a certain extent, be masters of our own actions—in spite of what some geneticists try to suggest.

### Yet it remains tempting...

Geneticists do not give up easily. They keep searching for genes that explain certain human features. The mantra of many geneticists is that there are genes for everything we do because they assume everything is programmed in our genes. Can that be true?

A few decades ago, the general estimate for the number of human genes was thought to be well over 100,000, but then turned out to be around 22,500 genes—which is only a little bit more than the 19,735 genes a tiny roundworm needs to manufacture its utter simplicity. And human beings have only 300 genes that are not found in mice. No wonder that the president of Celera, a biocorporation, said about this surprising finding, "This tells me genes cannot possibly explain all of what makes us what we are." At least, we have a first indication here that genes are not as almighty as some want us to believe. They cover at least the genes needed for metabolism—but there must be much more to a human being.

In science, discoveries always start as inventions—usually called hypotheses. However, not all inventions lead to discoveries. To use an analogy, the person who invented "Atlantis" (Plato, perhaps) did not discover Atlantis; it remains a legendary island until further notice. The same in science: Most inventions do not lead to discoveries. The Nobel laureate Peter Medawar's wise advice to a (young) scientist is that "the intensity of the conviction that a hypothesis is true has no bearing on whether it is true or not." Yet some scientists think they have made a discovery when all they have in mind is an invention, a hypothesis. Consequently, we have been bombarded with new genes: a gene for longevity, a gene for bisexuality, a gene for pedophilia, even a gene for religion—the list could go on and on. We could even invent a "chip gene" for people with an addiction to chips, a "chocolate gene" for chocoholics, or a "spending gene" for habitual big spenders.

The hypothetical genes some have come up with were once claimed, and then had to be retracted—they were often inventions that did not lead to discoveries. Apparently, genetic determinism is still rampant in genetics, hence hypothetical genes just keep coming and going. I am not declaring all these inventions bogus, but most of them are still in the stage of invention and are awaiting the stage of discovery. It just remains tempting to claim a genetic difference for something that may not even exist. Perhaps alcoholism is not entirely genetic but rather something acquired at home or in the womb or in a group of peers. Perhaps pedophilia is not an issue of genetics but rather a form of immoral behavior—rape, that is. Perhaps Munchausen Syndrome is only a call for attention or sympathy.

I am certainly not trying to put all these examples in the same basket, but I just want to signal how often we get bombarded with a new "disease." They invent a disease first, then they invent the gene to explain such a disease—and then they sell us an elixir to cure it! It may very well turn out that we have been dealing with phantom diseases, bogus genes, and useless elixers.

*Some pitfalls*

Let me illustrate the many pitfalls genetic research could be in for with the following case study. The American human geneticist Dean Hamer and his team once postulated a "gay gene." First, I must admit that Hamer did not claim that the gene itself causes male homosexuality, but rather its variants that may have an influence on the outcome. He also was aware of the fact that this was not "the" gay gene but rather one of potentially many whose vari-

## Chapter 9: Is Everything in the Genes?

ants might influence predisposition to male homosexuality. But most of those caveats did not reach the mass media.

The first challenge for Hamer's team was to localize this hypothetical gene on one of the 23 chromosomes we carry. They found a possible candidate when they asked 76 gay men about their relatives. It turned out that male homosexuality occurred more often among uncles and cousins on the maternal side than on the paternal side. All 655 relatives were asked to answer Kinsey's questions about sexuality. Although only 30 (!) lists were returned, there was a significant difference between the maternal and paternal side. Apparently—assuming both sides answered with equal honesty—the hypothetical gene is located on a chromosome that sons receive from their mothers, not their fathers. So, the search was focusing in on the X-chromosome.

Then they used so-called "genetic markers" on the X-chromosome—such as genes for hemophilia and color blindness which had a known position on that chromosome. We know that the chance of exchange (recombination) between two genes is proportional to the distance between them. In the study under investigation, recombination can only be recorded if we have at least two gay brothers in each family and preferably also their mother. This requirement was only fulfilled in 40 families (but, for some unknown reason, only for 16 mothers). If two gay brothers share not only their assumed gene for homosexuality but also a particular genetic marker, then that marker must be close to the gene we are trying to locate. Well, this happened to be the case for five markers at the end of the long arm of the X-chromosome. From the 40 pairs of brothers, 33 pairs had those five markers in common (13 more than the 20 to be expected by mere chance).

First of all, based on Hamer's research, all we know is that there might be some gene involved, but we do not know what the effect was of other genes and the environment. To find out how strong a gene's impact is, we would need research on adopted identical twins as well (more on this later). We do know from a 1991 study of identical twins that 52% (29 pairs out of 56) of these identical twins were both homosexuals; 22% (12 pairs out of 54) of the fraternal twins were both homosexuals; and 11% (6 of 57) of the adoptive brothers where both homosexuals. If homosexuality is genetically determined, why did only 52% of the identical twins share the same sexual orientation? How about the other 48% of the twins who differed in their sexual orientation? These are serious questions that need to be addressed, before claiming a "gay gene."

As a matter of fact, in a more recent 2000 study, the same researchers used volunteers, this time not recruited from the gay community, but from the Australian Twin Registry. They found that only 20%, and not 52%, of identical twins shared the same homosexual orientation. Obviously, it is very difficult to distinguish the genetic impact from the environmental influences on sexual orientation. In short, genes may have some impact but certainly not a monopoly.

One more caveat. Hamer's hypothetical gene on the X-chromosome may not be a "gay gene," but a gene that affects excessive maternal care by the mother instead. Did Hamer not say that mothers give this gene to their homosexual sons? Well, could it be that mothers with this gene "create" homosexuals? Psychoanalysts might tend to think so. I am not saying they are right, but I do say that there is no very strong evidence here at all. A mother's desire for having a boy may affect sexual orientation. All too often similar hypothetical genes—for example, genes for schizophrenia, autism, and bipolar disorder—were claimed, and then had to be retracted. They were inventions that did not lead to discoveries. Hypothetical genes just come and go in genetics.

*One more case*

Interestingly enough, the geneticist of a "gay gene" also came up with a "god gene." Hamer gave his 2005 book the provocative title *The God Gene: How Faith Is Hardwired into Our Genes*. The god-gene hypothesis was basically invented by Hamer, who now claims he has in fact discovered a gene that he decided to call the "god gene." To be more precise, he is talking about a gene for spirituality, which he deceptively dubbed as a "god gene."

Hamer theorized that if our sense of spirituality has a genetic basis, then those who rank higher in spirituality should share some genetic link that those who ranked lower do not. What did he mean by "spirituality"? He measured it by using a "self-transcendence" scale developed by the psychologist Robert Cloninger, in order to quantify how "spiritual" someone is, assuming that spirituality can be quantified by psychometric measurements. What impressed Hamer is that this self-transcendence measure had been shown to be heritable by classical twin studies.

What about the term "self-transcendence"? It is a word used by psychologists to describe spiritual feelings that are independent of what they call "traditional religion." Hence it is not based on belief in God, frequency of prayer, or any other conventional religious practice. Self-transcendent people tend to see everything, including themselves, as part of one great totality. They have a strong sense of "oneness" with people, places, and things. Self-transcendent individuals are also considered "mystical." They are fascinated with things that cannot be explained by science. They are creative but may also be prone to psychosis. In short, they are "spiritual," if you redefine this term in a certain way.

In order to identify some of the specific genes that might be involved in self-transcendence, Hamer analyzed DNA and personality score data from over thousand individuals. He asked them to fill out a detailed questionnaire—a standard test called a "Temperament and Character Inventory"—including a section that asked them to rate their feelings of "absentmindedness, connectedness with nature, belief in extrasensory perception, and other traits." He assumed that the answers would provide a measure of the subjects' affinity for what he called spirituality.

Then he went poking around in their genes to see if he could find the DNA responsible for their differences. With over 35,000 genes and 3.2 billion chemical bases in the human genome, he limited his search for the "spiritual gene" to nine genes known to produce monoamines (brain chemicals that regulate mood and motor control) and then identified one particular gene, VMAT2, as showing a significant correlation with affinity for spirituality. VMAT2 is a gene that codes for a monoamine transporter and plays a key role in regulating the levels of the brain chemicals serotonin, dopamine, and norepinephrine. These monoamine transmitters are in turn postulated to play an important role in regulating the brain activities associated with mystic and spiritual experiences.

When he analyzed this gene further, he discovered that those with the nucleic acid cytosine in one particular spot on the gene ranked high in spirituality, whereas those with the nucleic acid adenine in the same spot ranked lower. So, he concluded that a single change in a single base in the middle of the "god gene"— at position 33050 of the human genome map, to be precise—seemed directly related to the ability to feel self-transcendence. He even gave an explanation as to why the "spiritual" allele for this gene would give its carrier a selective advantage: Spiritual individuals are believed to be favored by natural selection because they are provided with an innate sense of optimism, which produces positive effects at either a physical or psychological level.

What are we to make of all of this? Let me mention first that Hamer rushed into print with his book without any peer review and without publishing his results in a credible and reputable scientific journal. Plus, which is even more serious, his findings have not been replicated. As someone said, "Given the fate of Hamer's so-called gay gene, it is strange to see him so impatient to trumpet the discovery of his God gene." All of this probably gives us ample reason to not take his work on face value—or to put it nicely, it definitely deserves further scrutiny. Here are some crucial remarks.

#1 The assumed link between gene VMAT2 and religious experiences is rather weak. First of all, why did Hamer limit his search to nine genes known to produce monoamines? If a certain gene variant is indeed the cause of being "spiritual," we should expect that the number of people possessing this variant should at least be proportionate to those who consider themselves spiritual. In addition, all those possessing the "right" variant should have spiritual experiences, otherwise the presence of cytosine cannot be the cause of being or feeling spiritual. Hamer failed to test for any ramifications like these. If our belief in the divine is due to our genetic wiring, how can one not believe in God when we are supposed to be "hard-wired" for religion?

#2 The results of twin studies only speak of some 40% to 50% heritability of spirituality and/or religion. This naturally raises the question of what to make of the residual percentages. Twin studies remain very controversial, even when done with identical twins who were separated by adoption. First of all, similarities between identical twins are not only the result of their identical genes but also of nearly identical surroundings. Their strong resemblances make it even more likely that others will treat them the same way in life. Moreover, they themselves often strongly desire to become and be more like each other. Hence, we would easily over-estimate the impact of genes.

Second, even if twins become separated at an early age, we need to take into account that any similarities between them get reinforced the longer it takes before they become separated. Let's not forget that, for nine months, they shared their mother's womb, including her voice, her hormones, her food, and her emotions. Besides, adoption usually takes place in an environment that is very similar to the original one, often just around the block or with relatives or friends. So, again, we tend to easily over-estimate the impact of genes. All of this makes research on identical twins, even when they were separated by adoption, rather limited.

#3 Even if it were true that we are genetically hard-wired for religion, what could this possibly mean? Clearly, we are not hard-wired for a specific religion; there are more than 7,000 identified "varieties" of religion. "Born a Catholic" does not mean always a Catholic. Plenty are the cases of people who, later in life, decided to become atheists or chose the opposite by leaving atheism behind. But in any of these cases, the changes in religion were almost certainly not caused by genetic changes.

#4 We are dealing here with genes that are allegedly connected with behavioral traits. Such hypothetical genes are supposed to control traits with very complex and variable patterns of behavior. They make for very intricate similarities that come in many variations. Most of these hypothetical genes were once claimed, and then had to be retracted. So, if we cannot even link individual genes to personalities, how can we possibly link genes to religion? There are too many intervening factors affecting this presumably simplistic link—for example, other genes, environmental effects on gene expression, cultural factors, upbringing, and personal experiences.

#5 It is very doubtful whether all this genetic talk has in fact anything to do with religion taken as a belief in a Transcendent Being, God. Perhaps genetics can tell us something about mystical experiences, but the idea that people believe in God because of mystical experiences is silly. One need not feel anything, let alone have a mystical experience, to believe in the existence of God. Arguably, most individuals who believe in God have never experienced God in a mystical way. Quite a few believe in God, or reject God, for purely intellectual reasons. Others simply have an intuitive awareness of God's existence or were brought up with that belief.

#6 A "feeling of transcendence" is not necessarily a religious experience, and if Hamer is right, it is in fact merely a biological one. The monoamines involved in the feeling of self-transcendence are the same monoamines that are jumbled by ecstasy, LSD, and other mind-altering drugs. If the feeling of transcendence is indeed a biological experience rather than a religious experience, then studies performed on that experience only tell us something about biology, not religion. The question of God's existence remains a philosophical and religious issue, not a biological one. While biology can tell us a lot about human beings, it does not and cannot tell us anything about God.

#7 All of this prompts the question as to why Hamer wants to reduce religion and faith in God to something else, to something like spiritual experiences, and ultimately to genetic instructions. The answer can perhaps be found in some underlying philosophical assumption that he seems to entertain—the assumption that biology can fully explain everything in life, including religious faith, beliefs, and experiences. It is a deep-seated dream of certain scientists. The sociobiologists E. O. Wilson once triumphantly exclaimed, "[W]e have come to the crucial stage in the history of biology when religion itself is subject to the explanations of the natural sciences." If I had to give Hamer's gene a label, "hallucination gene" would probably be the best fit.

#8 According to Hamer's ideology, even those kinds of behavior that we think are our own choosing—lifestyle choices, moral decisions, religious beliefs, and the like—would require us to postulate a gene for whatever we decide, believe, or choose. We labeled this before as the doctrine of genetic determinism—a belief or conviction, not a scientific discovery. Because it is rather rampant in genetics, new hypothetical genes just keep coming and going. We could even invent a gene that makes one believe in the all-powerfulness of genetics.

Let's round up this chapter. Determinism is a doctrine that wants us to believe that everything is predetermined. But how could it make us believe so, given the fact that our beliefs would be as predestined as anything else in life? Why bother to debate free will and similar uniquely human features if people are already genetically determined to either believe in them or not? Therefore, we can still maintain that we are not at the mercy of our genes. Our genes are not our fate or destiny; they are like a hand of cards we are dealt, but we can play them differently. In other words, human behavior is arguably more often than not a matter of lifestyle choices and favored beliefs rather than the outcome of a set of genetic instructions.

We are not really puppets of our genes. No real puppet has the high-level capability to even pose the question whether humans are merely puppets. Then why do we find ourselves so often in situations where we seem to be puppets? Because we allow ourselves to be "puppetized." But that does not prove that we are fully at the mercy of our genes.

# Chapter 10

## From Sex to Gender

Human beings are *mortal* beings—they don't live forever. You don't have to be a biologist to know that. Yet, there is a way they can "live on"— in the lives of future generations, that is. They can do so by having children. This requires sexual reproduction. Again, you don't have to be a biologist to know that.

Sexual reproduction means that an organism needs another organism in order to reproduce. So apparently the human species is somehow split in two parts: one can mate with members of one "half," but not with those of the other "half." Put more technically, one "half" produces egg cells, the other "half" produces sperm cells—and the two cannot produce new life without each other.

### Sex

All human beings carry 23 pairs of chromosomes. All their genes are located on these chromosomes, so chromosomes come in pairs as well. However, there is one pair of chromosomes, called the sex chromosomes, that is "matched" in females (XX), but "unmatched" in males (XY). During conception, both parents contribute only a half set of their chromosomes each, one of each pair, so their child ends up again with 23 pairs. Depending on whether the father passed on his X- or his Y-chromosome, the child will be either male (XY) or female (XX). If the egg cell is fertilized by a Y-bearing sperm cell, the new organism is a boy (XY), but if it is fertilized by an X-bearing sperm cell, the new baby is a girl (XX). This means a baby's sex is basically established at the time of conception.

Apparently, the presence of the human Y-chromosome determines that the child will be a boy—not quite, though, it is the absence of a Y-chromosome that determines the child will be a girl. To be even more precise, it is not just the Y-chromosome itself that is the determining factor, but rather the SRY gene located on the Y-chromosome which produces a protein called "testis-determining factor" (TDF). The SRY gene acts as a master switch and is responsible for the development of an unborn baby into a male by initiating the testes development (whereas other genes on the Y-chromosome are mainly important for male fertility). When this SRY gene is present, early embryonic testes develop around the 10$^{th}$ week of pregnancy. In the absence of the SRY gene, ovaries develop instead.

In general, one could say that sex determination of a male depends on the testes, and in turn, testis differentiation depends on the Y-chromosome and on its SRY gene which produces the TDF protein. It is indeed true that sex determination in rare cases can lead to abnormalities. However, it needs to be stressed that the existence of abnormalities does not make them "normal," as little as neurological abnormalities become "normal" by their mere existence—they still deviate from normalcy. Talking about abnormalities only makes the discussion confusing.

# Chapter 10: From Sex to Gender

Typically, after fertilization, development is "steered" in one of two directions, either boy or girl, male or female. I use here so-called fetal ages, counted from the estimated date of conception (some use gestational ages instead, which only start counting at the first day of a woman's last period, which is 2 weeks after ovulation). Based on fetal ages, we find at the 10$^{th}$ week that the penis of the male is slightly larger than the clitoris of the female. At the 12$^{th}$ week, the male scrotum has formed from the tissue that becomes the labia major in the female. Finally, at the 34$^{th}$ week, the distinctive features of the genitalia of the two sexes are fully present. Therefore, at birth, the anatomical differences between male and female are practically unmistakable. When the baby is born, the obstetrician or midwife announces, "It's a boy" (M) or "It's a girl" (F)—and it will stay that way for the rest of their lives. This is like a genetic disposition, which will never change.

## *Gender*

The Y-chromosome influences not only the development of the reproductive system but also that of the nervous system. And this in turn leads to differences in desires, behaviors, and skills between men and women. But this is not just genetically determined. The Y-chromosome mainly determines which reproductive organs develop. But from birth, parents and society also play a role in the "steering" of the child's development. As soon as parents see or learn that their baby is a boy, the child is treated as a boy. Children who are male in the eyes of others and are treated as male by others come to see themselves as members of the male sex. And something analogous would apply to girls.

To differentiate between what is genetically determined and what is rather socially determined, it is common to make a distinction between sex and gender. Gender is based on a social difference between men and women and thus goes much further than sex, which is based on a biological difference between the two "halves" of humanity—"males" and "females." Gender also means that men and women differ from each other in terms of character, wishes, expectations, skills, social roles, and so on. Whereas sex is engrained in the chromosomes, gender is acquired, partly based on cultural expectations, partly on personal choices. To put it simply, sex is what you are biologically; gender is what you become socially. In other words, the terms "male" and "female" are sex categories, whereas the labels "masculine" and "feminine" are gender categories. Your gender may be partially "self-made," but your sex is what you were born with.

Once the newborn has been "declared" a boy or a girl on a traditional birth certificate, based on anatomical and sexual characteristics, new features develop that can make them either more masculine or more feminine. During their further development, the difference between boys and girls becomes much more than a difference in biological characteristics—namely, differences in behavioral traits, social roles, and cultural expectations that come with being a man or a woman in a particular society. Once parents know the sex of their child, they typically treat the child either as a boy or a girl (unless they settle on what they had hoped for). In

## Chapter 10: From Sex to Gender

other words, early on in human development, parents as well as society take on a "molding" role.

The term "gender" was introduced in 1955 to distinguish it from the term "sex." (By the way, unlike in English, some languages use the term gender to mark nouns according to whether they are masculine, feminine, or neuter.) But what has remained standing is that someone's sex is biologically determined. It determines whether someone is born as a boy or a girl—period.

However, in the discussion about sex and gender, people often think about "genetics" in terms of a simple DNA mechanism. Some think that the DNA contains a blueprint of a certain end product, for example a human being. But that's not how DNA works. Others think that the DNA contains a recipe for a certain production process, from which, for example, a human emerges. But that's not a good picture either. DNA determines neither the outcome of the process nor the direction of the process, except in very exceptional cases such as the determination of blood groups, which comes more or less directly from the DNA. It is rather the case that DNA contains instructions for both building material and building tools and that a certain "construction" is thus realized in a certain environment. Each step in this process depends on the structure of the organism at that time and the current environmental situation. In that sense, there are hardly any properties that are purely or entirely "genetically determined."

This problem also plays a role with regard to the gender issue. The possession of sex chromosomes is by definition a matter of genetics, whereas the production of sex hormones is a bit further away from DNA, as sex hormones are not a direct product of genes in the chromosomes, but they are made by enzymes that are the product of genes. And the sensitivity of the various organs to these hormones will also be influenced by their DNA, but there are all kinds of steps and interactions in between. As soon as we start talking about the effect of sex hormones on brain development—and by extension, on behaviors, expectations, and self-image—the number of steps between DNA and the end result only increases, and with it the effect of environmental factors that intervene at each step.

In short, gender is more of a social construct, whereas sex is a genetic given. This distinction between sex and gender is important to keep in mind. For example, when we talk about homosexuality, we are talking in terms of gender. Homosexuality is not based on a change in sex—of producing either egg cells or sperm cells—but is rather a matter of gender instead.

### *The gender myth*

Nowadays, there is a strong tendency to use the concepts of sex and gender as almost identical—which makes the discussion very murky. This tendency comes from a new ideology telling us that sex and gender are both considered an entirely social issue, in denial of the fact that sex is a biological and genetic issue. We are dealing here with a myth, based on an ideology. It makes some people even say they were born in the wrong body. This creates a

## Chapter 10: From Sex to Gender

disconnect between the sex of a person and the gender of that person.

As a matter of fact, an XY person is going to be a male, whether he likes it or not. And an XX person is going to be a female, whether she likes it or not. Those are facts that cannot be changed, regardless of what we would like the facts to be. The human free will may have an effect on whether we accept or reject this fact, but it cannot change that fact. As a consequence, a person whose biological identity is male cannot have a female identity; if he still thinks he does, it is only in his head as an imitation of the other sex. We do not have the freedom to be either or both or neither, depending on our mood.

You don't need to be a biologist to know that boys are different from girls, as little as you would need to be a veterinarian to know that cats are different from dogs. It is a physical, empirically verifiable reality that does not change simply because our beliefs or desires about it do. One can surgically change the genitalia and sex glands that a person was born with, but not the person's sex. The physician Carl Elliott once remarked that cultural and historical conditions have not just revealed transsexuals but may in fact be creating them.

Sex is not unique in this respect. Something similar can be said about skin color, which is also strongly regulated by genes. Just as transgendered men do not become women, or vice versa—for our sex is part of what we were born with—in a similar way do transracial individuals not become members of another race by perception, because their skin color, too, is a part of who they are and of what they were born with. A white woman declaring that she is "in fact" black is just as odd as a woman declaring that she is in fact a man. Desires cannot change facts, no matter what our beliefs tell us. There is a fundamental difference between fact and fiction.

I keep stressing that a person's sex is a biological issue, whereas gender a sociological issue. However, that distinction has become very blurred in the minds of many nowadays. According to this new outlook, sex is no longer a biological concept, but it has become a social construct, in the same way as gender is. People who defend this view have chosen to change their terminology. For instance, in 2013 the American Psychiatric Association (APA) changed the term "gender identity disorder" by replacing it with the label "gender dystrophia." It is no longer seen as a disorder but as a case of distress.

Renaming labels is not an uncommon strategy. For instance, until 1974, the American Psychiatric Association had considered homosexuality a mental illness or disorder. When, in 1970, gay activists protested outside the APA convention in San Francisco, the APA decided that homosexuality be declared "normal," although only 58% of the members who voted favored the change. What is noteworthy about this is that there was no new fact or set of facts that motivated this major change; instead, it was the power of the gay lobby that initiated the revision. The psychologist Philip Hickey worded it this way, "So all the people who had this terrible "illness" were "cured" overnight—by a vote!" Talking about science....

This may explain why the so-called gender-identity disorder is on the rise; it seems to be spreading like wildfire. Because of an increasing number of broken families and same-sex parents, some children may not have the right parent to identify with, and therefore their gender may not have a chance to line up with their sex. As the lawyer Joseph Backholm said about gender ideologists, "The irony is that a sex change itself reinforces the gender stereotypes they claim to be rejecting."

As to the sex-gender issue, the problem remains that a person whose biological identity is male cannot have a female gender identity; if he thinks he does, it is only in his head as an imitation of the other sex. It is worth noting that when two people with same-sex attraction live together, one of them usually plays the male role while the other plays the female role—but these roles are obviously social roles, not sexual roles, or perhaps gender roles at best. And, of course, natural reproduction is out of the question.

However, the fact persistently remains that gender does not and cannot replace or alter sex. To think differently destroys a person's identity as a man or a woman. It obscures the reality of sex differences, making us believe that we can manipulate sex differences entirely to our own liking. It gives us an alibi to reject the binary division of persons into two sexes, so that we can claim the freedom to be either or both or neither, depending on our mood.

A person's is a physical, empirically verifiable reality that does not change simply because our beliefs or desires do. Therefore, one can surgically change the genitalia and sex glands a person was born with, but not that person's sex. One cannot "re-invent" oneself that way, because one never "invented" oneself to begin with. We must conclude from this that gender does not replace sex, nor does it nullify sex. Gender is a social construct placed on a biological foundation. Those who obscure the difference between sex and gender do so by equating both to a merely social construct. Again, there is a difference between fact and fiction.

In 2012, Pope Benedict XVI connected an extreme version of the gender ideology with the late French philosopher Simone de Beauvoir who once said, "one is not born a woman, one becomes so." This denies the fact that one is born as a woman but may not think so. Taken this way, the pontiff declared the "gender" ideology as being a new philosophy of sexuality: "According to this philosophy, sex is no longer a given element of nature, that man has to accept and personally make sense of: it is a social role that we choose for ourselves, while in the past it was chosen for us by society." The pontiff was basically asserting that this version of gender theory confuses sex with gender, and thus confuses sexual identity with gender identity. No wonder, in a 2023 interview with an Argentine newspaper, Pope Francis called gender ideology "one of the most dangerous ideological colonizations" in the world today. He added, "All humanity is the tension of differences. It is to grow through the tension of differences."

### *The mutilation*

The "gender" ideology has even led to medical and surgical interventions. However, receiving hormones of the opposite sex or having sex organs surgically removed, changed, or replaced are not sufficient to change a person's sex. They may change appearances, but not the sexual identity of a person. Nevertheless, there have always been people who tried to do just that. In the past, some used cross-dressing, but that has become passé with the advent of sex-change surgery. Now former cross-dressers can truly transform their appearances to an extent that clothes, wigs, and makeup never could achieve. But the idea behind all these trials is the same: changing appearances, and appearances only, can only be achieved through travesty and mutilation.

In 1979, after commissioning a study of the outcomes of sex-change operations, Dr. Paul McHugh in his capacity as chair of the Department of Psychiatry, put a halt to transsexual surgery at Johns Hopkins Hospital. He wrote, "We psychiatrists, I thought, would do better to concentrate on trying to fix their minds and not their genitalia." McHugh compares medical treatment of patients who have a confused gender identity to treating anorexia with liposuction. He calls transgendered individuals "feminized men or masculinized women, counterfeits or impersonators of the sex with which they 'identify.'"

A 2016 report by Dr. Lawrence Mayer and Dr. Paul McHugh, based on nearly 200 peer-reviewed studies of sexual orientation and gender identity, discloses some striking and alarming information. First, only a minority of children who express gender-atypical thoughts or behavior will continue to do so further into adolescence or adulthood. Second, among transgender individuals, 41% have attempted suicide, whereas only 4.6% of the overall U.S. population reports "a lifetime suicide attempt." The hypothesis of "social stress" as an explanation has so far not been corroborated.

Third, one hospital's practice of surgically removing the poorly developed genitalia of male infants and giving them female genitalia showed that, years later, most of the subjects still identified as male, although their parents had been directed to raise the boys as girls. Sometimes, or often, or always, ideologies ignore the facts. They are more concerned about their own narrative than about the truth.

More and more experts are beginning to realize, or at least acknowledge, what the gender ideology has done to the young generation. Now it has gone from genderism to transgenderism. Robert Royal describes it accurately as recycling "an old Gnostic heresy, wherein some inner reality, unconnected with the physical body, defines its own identity." Can that be done without serious repercussions?

Dr. Susan Jane Bradley, a Canadian psychiatrist and pioneer in the field of puberty blockers, now admits, "We were wrong," for the trans medical industry is harming children. She said about these puberty blockers, "They're not as reversible as we always thought, and they have longer term effects on kids' growth and development, including making them sterile and quite a number of things affecting their bone growth." She has also argued that gender identity disorder in children is sometimes rooted in serious family problems. That is the other side of the coin.

# Chapter 11

## Can Animals Talk?

Of course, animals can talk. They have been doing so for centuries, at least since the time when Aesop (c. 620–564 BC) wrote his famous Fables. And after that, more and more candidates have been added such as Reynard the Fox in the Middle Ages. And currently, we know of Bugs Bunny, Kermit the Frog, Winnie-the-Pooh, Donald Duck, Mickey Mouse and Minnie Mouse, Porky Pig, and Yogi Bear. Apparently, talking animals are a given in our world.

But, as you might say, these characters are fictional and not real. Nevertheless, don't be surprised that these myths have also started to take hold in science recently. A growing number of biologists, especially the behavioral biologists among them, are now letting animals do their "talking." They are in search of the talking animal. That should not surprise us, by the way, because since Charles Darwin, the gap between humans and animals has been narrowed drastically. This means, first of all, that Man has been seen as no more than a glorified ape, but also that the ape is nothing less than a Man-in-the-making. Humans are masters of anthropomorphism, so they tend to think that animals have got to be like humans, even when the differences are quite obvious. Apes are starting to look more and more like us, so they must be able to also talk a bit.

Well, do they really talk? It's hard to keep up a conversation with apes, you might say. But scientists will note that this could very well be a wrong impression. That is why behavioral biologists have tried to teach apes to talk under "experimental" conditions. Unfortunately, the structure of their vocal cords and larynx makes it impossible for them to produce sounds that resemble human speech. No problem, those scientists said, and that is why they turned to sign language, as used by the hearing-impaired, or to an artificial language that works with symbols. In one of these two ways, a young chimpanzee or gorilla was taught a form of "vocabulary" in a kind of family setting by researchers such as Francine Patterson and the Gardner duo, Allan and Beatrix. And indeed, by using gestures or symbols, these apes appeared to be able to "say" something.

They could also make "sentences" with several "words," although we must admit that the number of "words" per sentence is very limited, and there are no grammatical rules to be found. The climax, if I may use that word, was in the classic 1932 film *Tarzan the Ape Man* featuring the well-known dialogue between Jane and Tarzan with the immortal words spoken by Tarzan, "Me Tarzan, you Jane."

These are quite some achievements and claims. Do they truly stand up to scrutiny? Are animals really able to talk to us and to each other? It depends on what we mean when we speak of "talking." There are several possibilities.

# Chapter 11: Can Animals Talk?

## *Using signals*

If we equate "talking" to exchanging *signals*, then animals are indeed able to talk and communicate with each other. Many will point out that some form of "language" is readily available in the animal world, even among organisms much simpler than apes. After all, language is a form of information transfer. And don't we all transfer information, whether we are human, ape, bird, or insect?

The case seems to be quite simple. To transfer information, you only need signals. Living nature is full of signals. Just listen to the many birds in your area on a spring morning. There is quite some "talking" going on with all those chirping signals. But communication can also take place in the deepest silence. Honeybees returning to their nest make a nice cute dance back in the nest to "tell" their nest mates where to find food. If that's not information transfer?

In short, nature has developed a multitude of signals in the course of evolution. Such signals can be of various kinds—sound is just one of them. They have been developed in evolution to report something to other organisms, and that is why we speak of communicative signals—their function is communication. There are also signals not used for communication: for example, a predator flying in the sky can be a warning signal for other animals to hide, but it is not a communicative signal. In the case of a communicative signal, we should rather think of an alarm shout or something similar meant to alert other animals. Alarm signals have been selected in evolution for their survival value.

It is therefore not surprising that people easily speak of "the language of animals." However, although dogs can bark in different ways and birds can sing with a rich repertoire of songs, none of them are able to speak a language—perhaps they make sounds like we do, but that is not the same as using language like we do. Even parrots may sometimes sound like us, but they are not using language like we do—they merely imitate the sounds of human speech, without knowing what those sounds mean. Yet, we dare to speak of a bee language, similar to the way we speak of human language. As a matter of fact, the similarities are striking. Just as we can make sentences with words in human language, honeybees can use their bee language by varying the angle and durations of their dances to inform other bees. Dancing is "the message" here.

Signals are best known as powerful warning tools. To indicate specific dangers, some animals have even developed an extensive repertoire of alarm signals. For example, we know about green meerkats that they warn their fellow monkeys with a barking sound as soon as they spot a leopard, but with a coughing sound when an eagle is overhead. Different signals are responded to in a different way. Even when these sounds are broadcast through a hidden speaker, these monkeys respond to a barking sound by fleeing into the trees, and to a coughing sound by running into the bushes. They then do not respond to the enemy itself, but to a signal associated with that enemy. Are they "talking"? In a sense they are.

## *Using labels*

If we equate "talking" to using *labels*, then animals may also be able to talk. It is quite obvious that animals, and apes in particular, do have the ability to classify things in categories—for example, the category of what is dangerous versus what is not. That ability is of vital importance, because it is often best to react in the same way to the same things, while reacting differently to different things. In other words, whether things are the same, whether things should be distinguished, depends on what is best for responding to them.

Classification is vital for behavior. It allows animals to "label" things in their surroundings. They have somehow "labels" to mark what is dangerous, what is edible, who is a potential partner, and so on. So, the ability to classify things must be present in the animal world. In short, survival in a world as we know it requires some kind of anticipation and planning—not just for humans, but also for animals. In this sense, every organism must be able to classify and generalize situations.

No wonder then, scientists have worked on the ability to classify things. The more recent heroes in our "pantheon" of talking animals are the world famous chimpanzees Sarah and Washoe, as well as the equally amazing "linguist" Koko the Talking Gorilla, and more recently the amazing Nim Chimpsky, the Talking Chimpanzee. The late David Premack, Professor of Psychology at the University of Pennsylvania, was the one teaching Sarah the Chimp to communicate by using cards. And the Gardner duo, Allan and Beatrix, taught Washoe the Chimp to use gestures inspired by the

American Sign Language. And Francine Patterson taught Koko the gorilla to understand more than 1,000 signs of what she calls "Gorilla Sign Language." Also Herbert Terrace and his team made the chimp Nim learn 125 signs of American Sign Language.

In all these cases, the words taught to these apes were in fact "labels." Labels can be powerful. Thanks to such labels, the apes mentioned above were not only able to designate things with a label, but also group different things under the same label. For example, pears, apples, and bananas were in their "language" all "fruits." But sometimes, the "grouping" goes to extremes. Take the fact that the chimp Nim was able to use the label "apple" not only for apples, but he used the same label "apple" also to refer to the action of eating apples, to the location where apples are kept, and even to events and locations of objects other than apples that happened to be stored with an apple (the knife used to cut it). In other words, the label "apple" was used by Nim to refer to anything associated with apples.

Apparently, the use of labels is not only powerful but also quite limited. Although the chimp Nim was able to learn 125 signs of American Sign Language, he never got beyond memorized two-word combinations without any syntactical structure. They were merely labels, but not language. The linguists Robert Berwick and Noam Chomsky say about Nim, "All that Nim was actually able to learn about American Sign Language was a kind of rote memorization—(short) linear sign sequences. He never progressed to the point of producing embedded, clearly hierarchically structured sentences, which every normal child by age three or four can do."

Chapter 11: Can Animals Talk?

## *Using words*

If we take "talking" in the sense of using *words*, then the discussion enters an entirely different territory. Why is that? Well, for at least two reasons.

Reason #1 is that words are more than *signals*. Let me explain this with a simple example. When I utter the word "boss" to a dog on the street, the animal won't sink into deep thoughts, but will (if properly trained) react as if the boss is nearby. This dog won't take "boss" as a word that evokes a certain thought, but merely as a signal that calls for a specific behavior.

For humans, that may work the same way. They can use many words as mere signals—for example, when I exclaim "The Boss!" and then everyone quickly goes back to work. But the word "boss" is also more than a signal. It has a meaning that we use even when the actual boss is not around. Sometimes the word refers to a particular boss, but usually to an abstract idea we have in mind. Words have their own specific meaning, independent of what they may refer to. This difference separates the world of animals from the world of humans. Animals treat everything in their surroundings as signals that call for a direct response (association), but they cannot use words to ponder realities beyond their needs for food and sex. Signals are situation-specific, whereas words usually are not. Signals call for direct action, whereas words typically do not.

Reason #2 is that words are more than *labels*. Take, for instance, a word such as "poison." Indeed, it can be a label placed on a bottle to warn people of danger. But it also tells us something about the nature and working of what is inside the bottle. In other words, the word "poison" is more than a label of classification. It comes with a "thought" hidden behind that term.

What is it then that makes words so different from signals and labels? Well, words have a *meaning*, which signals and labels lack. This takes us into the field of semantics. Words are elements with a meaning, hidden in the *concept* behind the word. It is the concept that relates the word to the world of abstract ideas and thoughts, more than to the world of concrete entities. Words can even refer to something that is not present anywhere, or to something that does not even exist. The word "centaur," for instance, does have a concept behind it, but does not refer to any real entity in the world. That is why we can also use words to tell stories about talking animals!

True, words can be *used* as signals or labels, but they go far beyond them. They may not even refer to anything concrete in the world around us. To explain the concept "tomorrow," for instance, there is nothing to point at. To explain the concept "circle," there is nothing in the world to point at, for nothing in the world has the perfection of a circle. Instead, this concept is based on other concepts such as radius and diameter. Or take the words "red" and "blue" which are based on the concept "color" (with a certain wave frequency). Concepts have a universality that objects in the world can never have—the concept "circle," for instance, applies to every possible circle without exception.

Language users can only use a word correctly if they have mastered the underlying concept (the "meaning"). That's why people can sometimes use words without really knowing the concept behind them. And that's also the reason why animals can use words without knowing there is a concept behind them, so they are merely using them as signals or labels. Animals do have the ability to categorize things, but that does not imply that they also have the ability to use concepts. Indeed, using concepts implies the ability to categorize, but one cannot conclude from this that the ability to categorize implies the ability of using concepts—that would be a logical error. Hence, it does not follow that animals use concepts—that is, words with a meaning—in addition to signals and labels that evoke action. Words have a meaning that signals and labels lack. That meaning connects them with the world of thoughts.

*Using sentences*

If we equate "talking" to using *sentences*, then the gap between the human world and the animal world becomes even wider, perhaps even to the point of being unbridgeable.

The fact of the matter is that talking is more than uttering a series of words. Words are part of larger entities such as sentences. Every language has rules that determine how words can be linked together. Take the word "boss" again. It is a signal that may sometimes refer to a particular individual. Yet, it is in essence a word for what we think and know about any boss in general. But—and this is an essential addition—the word "boss" is also a *noun*. This calls up certain combination possibilities.

The word "boss" may be preceded by an article: *the* boss. An adjective can be placed between the article and the noun: the *strict* boss. An addition can be placed after the noun: the strict boss *of Mary*. These options are completely independent of the underlying concept. Instead, they are determined solely by the fact that the word "boss" is a noun. The underlying concept and the intended reference only determine which grammatical combinations have meaning and which are true. But it is the grammar that determines which combinations are possible.

In addition, words can be combined into groups, which in turn can be combined with other groups. And we can even repeat this endlessly: "John's old hatred for Mary's strict boss." This endless combination power of human language can be found in any of the more than 5,000 languages spoken by humans. It is, in other words, a universal feature of human language. In human language it is very important where exactly a word is located. For instance, according to the sentence "John is washing himself" it is John who is washing himself, but not in the sentence "John's cat is washing himself."

What then about apes? True, they can connect "words" with things, but no connection to a grammatical combination system has ever been found. Even when they are taught a symbolic language, they lack what people would call a "language ability." And with that, the language ability of people turns out to be something unique. It's striking to note how humans almost "instinctively" use language, whereas apes do not. Apes must be forced to "say something," whereas children do develop this capacity naturally (otherwise we would call upon a therapist for help). *Homo sapiens* is "by nature" also a *Homo loquens*, equipped with a unique language capacity. A new-born human infant instantly selects from the environment language-related data, whereas an ape, with the same auditory system, only picks up noise.

Experts in the field of linguistics have pointed out that the simple "sentences" apes like Washoe, Koko, and Nim seem to form turn out to lack any kind of grammar—they are something like "me Tarzan, you Jane." In other words, language is more than a communication tool by using labels connected to objects. Words in human language are not just "labels," let alone "signals." True, animals may very well be able to use labels—as Sarah the chimp did with plastic tokens—but whether they can use language is a completely different issue. Apes may be able to link labels to objects, but they have never been able to link them to other words in a grammatical, recursive, and structured way.

Apparently, we have here another unique feature of human language, its *grammar*. Even Premack himself, the "grand old man" of behavioral studies on Primates, eventually changed his originally positive views about Sarah's language abilities. After more than 25 years of research on the origin of language in the animal world, Premack had to admit that the emergence of human language is, in his own words, "an embarrassment for evolutionary theory."

## Anthropomorphism

However, many behavioral scientists don't give up easily. They tell us that apes have the gift of convincing each other, but also of cheating; that they can interact with each other, and so on. If that isn't proof of them having thoughts, they exclaim. Apes are seen as "born psychologists" able to attribute expectations and intentions to each other, but also to other animals or people. For example, Premack trained chimpanzees to study video tapes in which a person does something that goes wrong. Then they were shown recordings where the act continues; only one of the tapes shows the person correcting the error. Since chimpanzees can identify that one shot, Premack says that would be an indication that they have an inkling of what the person being filmed had in mind to do.

All this seems to indicate that chimpanzees do have *thoughts* and can even attribute thoughts to others. They can communicate and act based on mental representations about the mental representations of others. We could express this process in words like these: "I think that you think that..." It is a thought about another person's thought. Isn't that a form of language?

On the other hand, one could object that this conclusion goes too far and is not a thought about a thought, but just an anticipation of how someone else might behave. Nevertheless, it seems obvious that there is a thought involved here. In order to lie and cheat—which monkeys and apes are well able to do—you have to be able to respond to the thoughts of other animals. And where thoughts are involved, we could teach apes to put their thoughts

## Chapter 11: Can Animals Talk?

into words. In principle, monkey language and human language could then be almost identical.

Do animals really have thoughts? The problem for them is that they cannot tell us about their thoughts. Yet there is at least one way animals could reveal their presumed world of thoughts—and that is through language. But what if they don't have option. People have become masters at using language. They can transmit mental images that they have often already received from others and then perhaps tested, modified, and expanded themselves. Language enables people to stand on the intellectual shoulders of their predecessors.

Language is the information carrier, but not the information itself. The information carrier "language" works with sentences that are built up from words according to grammatical rules—and by using them a certain message or information can be conveyed. The message itself, on the other hand, consists of certain concepts, mental representations and reasonings. The ability to speak therefore presupposes the ability to think. Certain cognitive abilities, such as the ability to classify and to reason, are the foundations on which the language is built. But there is much more to it: as we found out earlier, without a mind that harbors concepts there is no language.

What then does language have to add to thinking? Perhaps it is an aid to structure or refine our thinking. Moreover, language has a so-called iterative capacity, which means that the sentence "I think P" can be extended to "I think you think P." etc. In this way, thoughts about thoughts can be easily expressed. But more importantly, language is an information carrier that helps us transfer information to each other. Animals can transfer behaviors to each other. When one animal performs a certain behavior, other animals can imitate it.

This brings us to the heart of the matter. Since animals have little or no real language at their disposal, and since language is the pre-eminent means of conveying a world of thoughts to others, animals have little or no way, even if they did have thoughts, to tell us about them. This explains the problem we are grappling with in this chapter! If animals have no thoughts, then they cannot have language. But if animals do have thoughts, then they cannot convey them to us without language.

Some philosophers and biologists say that we simply don't know if animals have thoughts because they can't talk about them. For example, the philosopher Quine says that when we assign expectations and intentions to animals, we put ourselves in the shoes of the animal, so to speak, and speak on behalf of the animal what we ourselves would think or say if we barked at a cat in the tree. Attributing thoughts to humans is a reasonable idea, on the grounds that humans express themselves in language in that sense—but for animals, this is at best "a manner of speaking."

On the basis of what we have seen above, animal behavior can be explained without assuming an underlying world of thoughts. To claim differently or to come up with talking animals, is an unwarranted form of anthropomorphism.

# Chapter 12

## Can Machines Think?

This question may sound baffling to many. No one would have asked that question centuries ago. But times have changed. Machines have become quite sophisticated—so sophisticated that a human being, too, might best be understood as a machine. It is an old idea that has become more appealing these days.

**The machine in Man**

It is mainly thanks to the philosopher René Descartes that we talk about the human body as if it were a machine. In 1637, he argued that the world is like a machine, its pieces like clockwork mechanisms, and that the machine could be understood by taking its pieces apart, studying them, and then putting them back together to see the larger picture again. It is that nuts-and-bolts approach that he applied also to human beings. He once wrote, "I should like you to consider that these functions (including passion, memory, and imagination) follow from the mere arrangement of the machine's organs every bit as naturally as the movements of a clock or other automaton follow from the arrangement of its counterweights and wheels."

The idea that animals and humans are completely mechanistic automata that just follow all the physical laws of the universe and are controlled by the machinery of their bodies, was soon used by the 18th-century French physician and philosopher Julien Offray de La Mettrie in a 1747 book that he gave the provocative title *Man a Machine* (in French: L'homme Machine). This book, however, offered an extreme version of Descartes' idea, suggesting there is no difference between living matter and dead matter, thus eliminating the notion of a soul, which Descartes still had safeguarded.

The machine metaphor for human beings has never left us since. Understanding man as a machine has in fact been very successful. We think of the heart as a pump that processes blood, the eye as a camera that processes images, and the stomach as a concrete mixer that churns and mixes food with gastric juice. So, it is probably no coincidence that many discoveries about the working of the human body were inspired by the latest technological contraptions of the time. Our understanding of many human organs came in the form of machine-like mechanisms. The camera with its lens helped us understand the working of the human eye. Bellows clarified how the lungs can do their work. Pumps revealed what the heart does for blood circulation. In line with this thinking, food does for the body what fuel does for the steam engine.

No wonder then, when things go wrong, we use technological devices to correct them: prescription lenses, hearing aids, pacemakers, and prostheses. Even in healthcare, we think of a human being as a machine-like mechanism that needs to be "repaired" if it isn't working properly, so we can fix or replace parts of it as if they were parts of a machine. In all these examples, the case could be made that technology has been an important driving force for scientific advancement. It has been said that science has profited more from the steam engine than the steam engine from science. As a matter of fact, scientific discoveries often start with technological inventions.

The point is that an organism can be treated as a machine, as if it were an intricate network of "cogs and wheels," with one cause being set into motion by another cause, thus working with clockwork precision. Nowadays, this quite crude mechanical model of cogs and wheels has been replaced, or at least expanded, by the biochemical model of enzymatic reactions—yet the outcome is the same. Through this approach, the human body has become a "machinery" of biochemical pathways. Now the brain is supposed to secrete thoughts the way the pancreas secretes insulin. This is how Descartes and his followers taught us to "see" what man is like. Since then, we almost automatically look for "the machine in Man." And that approach has not been without success, because we have come to see many life phenomena more clearly through the imagery of a machine. It makes for a "nuts-and-bolts" approach of the human body.

Let me make clear from the outset that there is nothing wrong with studying an organism as a machine—the success of the machinery image proves it. But this harmless technique becomes dangerous when we claim next that an organism *is* in fact a machine, and nothing more. However, looking at an organism as if it is a machine does not *make* it a machine.

Again, the "model" of a machine can be a helpful tool to better understand the working of an organism, but it remains a model.

Models simplify what is considered complex, and thus they make complexity more accessible, controllable, and manageable. However, this harmless technique becomes dangerous when we claim next that an organism *is* in fact nothing but a machine. All models are merely simplifications of reality. They are never an exact replica of what they represent—the only exact replica of an organism would be that organism itself. A road map, for instance, is a model of the countryside, but you cannot drive on a map like you drive on a road. Put differently, each model depends on "how you look at it," for all science is based on abstraction. Using the model or metaphor of a machine, too, is a certain way of "looking" at the world in abstraction.

**The brain as a machine**

Not too long ago, another technological contraption, telephone connections, helped us understand how the nervous system works. But it didn't stop there. New machines have been developed that did not exist before. One of them is the computer. Perhaps computers can help us better understand how the brain works. Therefore, it should not come as a surprise that the human brain is now treated as if it were a computer. It seems that the computer can help us get rid of the last non-machine piece in man.

It is quite easy to see why this model is so attractive. We now know how the computer works and we know how the brain works. They both process information in a "digital" way. The brain does this from neuron to neuron according to the principle of whether or not to fire, while the computer does so from transistor to transistor according to the principle of whether or not there is a current. So, they both work on the basis of the same code—on/off. We call this a *binary* code, which formally consists of "ones" (1) and "zeros" (0). That is a formal designation for a large number of alternative states, such as on/off, empty/full, open/closed, high/low, up/down, left/right, yes/no, and so on.

So, it turns out that brains and computers both process information by means of a binary code. What happens in the brain is supposedly the same as what happens in the computer, so the claim goes! In other words, the brain thinks like a computer thinks. That conclusion seems inevitable. However, if that were true, would there still be any difference between "real" thoughts in the brain and "simulated" thoughts in the computer? For a moment you might still think that "real" is something different from "simulated." For example, it is possible to simulate the digestive process on a computer screen by manipulating images with ones and zeros according to a certain program. It now seems easy to argue that simulated food on a screen is as real as food in the intestines. Nevertheless, real food is made of different "material" than simulated food. So, in that sense, they are certainly not the same.

In the case of food this is clear, but when it comes to *information*, things are different according to some scientists and philosophers these days. "Real thoughts" are—they claim—based on the binary code of the brain and are therefore just ones and zeroes, just like simulated thoughts. But if that were true, then there would be no difference between real thoughts and simulated thoughts.

They would both be made of the same "material"—information, that is.

Indeed, information is always expressed in a certain code. A code is not a material thing but a formal pattern. "In-formare" originally means bringing the material into shape, or capturing the form in material. In other words, information is a certain material structure—for example, ink on paper, chalk on a blackboard, sound waves in the air, electric current in wires. The material used is the carrier of a code. It transports codes—and with it, information—like a car can transport people.

As a matter of fact, any code can be reduced to a binary pattern—for example, to pairs such as on/off (transistors and neurons), short/long (Morse), up/down (Braille), left/right (Abacus). Apparently, information can be encoded in different ways; the same thought can be expressed in Morse, but also in Braille. Consequently, it is not the information itself that is different but the carrier of the information.

But once we admit that the same information can be encoded in different ways, it is no longer possible to equate code and information. The code is not the information itself but is only a vehicle for information. The same information can be expressed in a binary code or in a polynuclear code such as DNA, because codes are only information carriers. Therefore, the fact that real thoughts and simulated thoughts have the same binary code does not mean that their information is identical. What real and simulated thoughts have in common is the same kind of code, but that's only one side of the information. Seen in terms of syntax, real and simulated thoughts may be the same, but that says nothing about the information they contain, which is rather a semantic issue.

According to the semantic approach, information has not only a code but also content and meaning—that is, information refers to something outside itself. In other words, information is *about* something. Information is more than a series of ones and zeroes. Without reference to anything beyond the language's symbols, any language would be a closed system. In a dictionary, words can only be explained with the help of each other; this vicious circle can only be broken if words are about something outside the language. It is therefore said that real thoughts are *intentional*: they are about something, and therefore language is typically also about something.

This takes us back to the difference between "real" thoughts and "simulated" thoughts. Well, the Berkeley philosopher John Searle argues that "real" thoughts in the brain are intentional, whereas simulated "thoughts" in the computer are not. So real thoughts would be more than a series of ones and zeros. Syntactically they may be structured the same as simulated thoughts, but semantically they are not the same. What computers simulate is just the syntactic aspect of information. Computers are at best syntactic, but not semantic entities.

To make his point, John Searle introduced what is widely known now as the "Chinese Room Argument"—first presented in 1983. Over the last five decades, this argument was the subject of many discussions. In January 1990, the popular monthly *Scientific American* took the debate to a general scientific audience.

Searle himself summarized the Chinese Room argument as follows. Imagine a native English speaker who knows no Chinese but is locked in a room full of boxes of Chinese symbols (a data base) together with a book of instructions for manipulating the symbols (the program). Imagine that people outside the room send him Chinese symbols which, unknown to the person in the room, are questions in Chinese (the input). And imagine that by following the instructions in the book, the man in the room is able to send back Chinese symbols which are correct answers to the questions (the output). Then, as Searle asserts, this program enables the person in the room to pass the so-called *Turing Test* for understanding Chinese, yet that person does not understand a word of Chinese.

The question Searle wants to answer is this: Does the machine literally "understand" Chinese? Or is it merely simulating the ability to understand Chinese? If you can carry on an intelligent conversation with an unknown partner, does this imply your statements are really "understood"? Searle argues that what the machine is doing cannot be described as "understanding," and therefore, the machine does not have a "mind" in anything like the normal sense of the word. A computer can translate language—as several translator tools online show us—but the computer does not really understand language, let alone think of itself as an "I" who does the understanding.

What Searle means by "understanding" is what some philosophers call "intentionality"—the property of being *about* something, of having content. It is the "mystery" of how mental thoughts about objects come to be about those very objects. Unquestionably, thoughts are about something beyond themselves. To use an analogy, anything that shows up on a computer monitor remains just an "empty" collection of "ones and zeros" that do not point beyond themselves until some kind of human interpretation gives sense and meaning to the code and interprets it as being "about" something. Think of what we call a picture: It may carry information, but the picture itself is just a piece of paper that makes "sense" only when human beings interpret the picture as being "about" something else. The same with books: They provide lots of information for "book worms," but to real worms they only have paper to offer. That's where the need for intentionality and about-ness comes in.

This raises the question as to whether a computer really "understands" what it's doing. Instead, it can be argued that it merely manipulates symbols or numbers that mean something to the human programmer. But do they mean anything to the computer, too? Does the computer "know" that the string of symbols it prints out refers to books, for instance, rather than to melons or planets?

Seen in this light, understanding information must be more than merely making use of that information, otherwise we would end up with bizarre conclusions. The physicist Stephen Barr makes the following comparison:

> *An ordinary door lock has "information" mechanically encoded within it that allows it to distinguish a key of the right shape from keys of other shapes. [...] Does the lock understand anything? Most sensible people would say not. The lock does not understand shapes any more than the fish un-*

*derstand shapes. Neither of them can understand a universal concept.*

### The brain works with concepts

There are those concepts again. As a matter of fact, universal concepts are uniquely human. Whereas the brain may be able to handle signals and images, it seems that only the human mind can deal with concepts. Images can have some degree of generality—we can visualize a circle without imagining any specific size—whereas concepts have a universality that images can never have—the concept "circle," for instance, applies to every possible circle without exception. Because images are inherently ambiguous, open to various interpretations, we need concepts to give them a specific interpretation. That's how mental concepts transform "things" of the world into "objects" of knowledge, thus enabling humans to see with their "mental eyes" what no physical eyes could ever see.

Concepts may be compared to searchlights which may bring to light what had escaped notice before because we had the wrong concepts. This does not mean, of course, that what we capture with concepts does not exist in the world around us but only exists inside our minds. We should think about concepts as a source of light in the way the philosopher Edmund Husserl sees it: If a light beam hits a certain thing that is in darkness, this thing will be in the light, and yet it would not be inside the source of light. In other words, these light beams do reveal to us what the world is like. When the physicist Isidore Rabi was told of the discovery of a new

# Chapter 12: Can Machines Think?

elementary particle, the muon (a sibling of the electron), he reacted with skepsis and replied, "Who ordered that?" But it soon turned out it was not just invented but also discovered to be a real subatomic particle. The invention has become a discovery.

A concept may be as simple as a "circle" or as complex as a "gene" or a "muon," but a concept always goes beyond what the senses provide. Concepts are not perceived images, as images are by nature ambiguous, open to various interpretations. Therefore, we need concepts to interpret what we perceive. We do not "see" muons but have come to hypothesize and conceptualize them with a concept. We do not even "see" circles, for a "circle" is a highly abstract, idealized concept (with a radius and diameter). We can even conceptualize something that we cannot visualize—something like a circle with four dimensions, for instance. Once these concepts have been established and mastered, we have become "regular observers" of "circles" and "genes" and "muons." But again, these are not images or pictures but concepts.

Now, the question that immediately arises is how information can acquire content. "By nature," information codes have no meaning. I was recently sent a poster with a series of 26 different types of butterfly wings, each one carrying the image of one of the letters A through Z. The fact that such a collection can be found in nature does not, of course, mean that our alphabet is a universal "code of nature." Greek, Hebrew, and cuneiform script, for example, use different codes. The pattern on the first wing of the poster may look like an "A" for many peoples, but not for all peoples, and certainly not for butterflies. Books—to take another example—are full of information for readers who understand the code, but they have only paper molecules to offer. In short, it's the user of the code that assigns information content to the information carrier. Information is the result of a triangular relationship between user, code, and content.

Because of this triangular relationship, computers can only provide information for users who understand the symbols of their output. In principle, everything—from calculations to fireworks—can be simulated on a computer, but the information content depends on the information user. Computers have no built-in semantics. Without semantic interpretation from the user, the computer has nothing to offer other than a collection of ones and zeros hidden in chips and pixels.

Perhaps we can go even further: without a user, the computer doesn't even have a syntax. There are no ones and zeros in a computer at all, but there are two alternative states that we humans *use* and *interpret* as a binary code of ones and zeros (1/0). In principle, those two states can be anything—on/off, empty/full, high/low, up/down, left/right, yes/no—and with that, there are actually countless forms of computer systems possible, although not all of them are equally efficient. We can choose from a wide range of systems—from an abacus to a light computer—as long as the physical properties of the system lend themselves to a useful syntactic interpretation. Without the user's syntactic interpretation, a computer is just an electronic circuit.

In short, the hardware of a computer does not even have syntax, let alone semantics. The material is the carrier of a code, and the code is the carrier of a meaning because both code and meaning are part of a communication system set up and designed by the users. Without user(s) there is no syntax and no semantics behind the binary code. Information only exists in the heads of the users.

*Artificial intelligence*

Computers can do more than carrying and transmitting information. They can also do other, very diverse things, such as recording and checking measurement data, editing photos and drawings, making music, or even controlling robots.

We are entering here the field of *Artificial Intelligence* (further abbreviated to AI). John McCarthy, who coined the term AI in 1955, defines it as "the science and engineering of making intelligent machines." That sounds quite harmless, but AI claims go much further than what a personal computer, or even a robot, can do. Its central claim is that a central property of humans—their intelligence—can be described so precisely that a machine can be built to simulate it.

There is no a priori reason why simulating human *intelligence* would not be possible. In general, intelligence works with perceiving sense-data and processing them. Many animals show some form of intelligence in their behavior, because intelligence is a brain feature and as such an important tool in survival. Therefore, animals can process images more or less intelligently. They can show various forms of intelligence: We find spatial intelligence in pigeons and bats, social intelligence in wolves and monkeys, formal intelligence in apes and dolphins, practical intelligence in rats and ravens, to name just a few. In other words, intelligence is a matter of processing sense-data—something AI systems could do as well if they are outfitted with the proper "sensors." Robots, for instance, can "cleverly" process sounds, images, and the like, and then react "intelligently." Their "intelligence" is not a uniquely human feature.

But sometimes, the claims of AI go much further—which is often called "strong AI." This extension was already behind some of the statements of early AI researchers. For example, in 1955, AI founder Herbert Simon declared that we are able now to explain "how a system composed of matter can have the properties of mind." Cognitive scientist John Haugeland went as far as writing that "we are, at root, computers ourselves." More specifically, one could state that strong AI, in David Cole's words, represents the view that suitably programmed computers can understand natural language and in fact have other mental capabilities similar to the humans whose abilities they mimic.

The keywords here are "mind" and "mental capabilities." Mental capacities of human beings include much more than intelligence. Why is that? Humans are not only outfitted with intelligence but also with *intellect*. Intellect enables us to know and understand things around us. It is vastly different from intelligence. Like intelligence, intellect can also use sense-data, but unlike intelligence, it changes perception into cognition. It does so by using mental concepts, which make sensorial experiences intelligible for the human mind. Whereas intelligence can be graded on an IQ scale, intellect cannot. One can have intelligence to different degrees, but one cannot have intellect to different degrees. Intelligence is a *brain* feature and helps us *navigate* the world. Intellect, on the other hand, is a *mind* feature and helps us *understand* the world.

We discussed already that concepts play a crucial role in how we know and understand the world. Thanks to concepts, we can see similarities that are not immediately evident through the senses because they are not directly tied to what we perceive. Unlike intelligence, the intellect changes perception into cognition by means of mental concepts, which make sensorial experiences intelligible for the human mind. Everyone can see things falling to the ground, but to perceive gravity one needs the concept of gravity in order to see what no one had been able to see before Isaac Newton. The concept of gravity allows us to "see," for example, the similarity between the motion of the moon and the fall of an apple. Again, that's not a matter of intelligence but of intellect. So, there may be artificial intelligence but not artificial intellect.

That's why "strong AI" claims go off track. If we take "intelligence" as more than processing information—something a robot can do as well—but also in the sense of the mind's capacity of "thinking" and "understanding," the question becomes: Would an appropriately programmed computer with the right inputs and outputs just simulate a mind, or actually have a mind with an *intellect*? The former part of the question, simulating a mind, is relatively neutral, but the latter part, literally *having* a mind, would have enormous philosophical repercussions, if it were true.

But there are other problems with the "strong AI" claims. They seem to imply that these machines in fact create themselves. But that claim wouldn't make sense, for nothing can cause itself and make itself exist. To say that a machine created itself is in fact philosophical nonsense, for then such a machine would have to exist before it came into existence. Calling the machine an "animated machine" makes the situation even worse, for where does the "animated" part come from? Calling the machine an "automated" machine evokes the same problem: where does the "automation" part come from? Machines, curiously enough, are always made for a purpose. The world of technology is per definition purpose-driven, based on purposes that designers and engineers have in mind.

Our criticism of "strong AI" could go even further. Machines only exist because they have a human maker. A computer can be programmed to make "decisions" with an if-then-else structure, for instance. It looks like "If X, do this" and "If Y, do that." But this does not mean that the computer actually deliberates and decides—programmers did, and they make it look like the machine does too. The "reason" why calculators calculate, or pumps pump, or computers make decisions is that a machine was designed for that purpose—and not because it "intends" to.

Machines don't have purposes, but humans do. In other words, computers only do what we ourselves—human beings with a mind and its intellect—cause them to do, for we have proven to be champion machine builders. Because of this, the existence of AI systems presumes the existence of a mind that designed them. The parts of a robot do not come together on their own and then function as a robot, but they must be arranged by a designing mind to do so.

To put this in a catchphrase, not only is there "a machine in the Man," but there is also "a Man in the machine"—the man or woman, that is, who designed the machine. So, the popular slogan "Man versus Machine" should actually read "Man versus Man"—versus the Man who built the machine. Are these considerations final and decisive? In a sense, they are because of logical and a priori considerations. Perhaps they are not, but they cannot just be ignored either. Only time will tell what AI fans are capable of creating, but that should not stop us from asking some critical questions guided by sound reasoning.

## *Man in the machine*

If the above considerations are correct, then there can be no thoughts in a computer. Because if we were to equate information from the computer with information from the brain, the question is who or what would then syntactically and semantically interpret the neuronal interactions in the brain—or in the computer, for that matter—so that it become information for the human mind with its intellect. Even if we were indeed "flesh and blood computers," then there must be "someone" in the brain behind the buttons of our neuronal computer to turn a binary code of alternative states into a code with content.

Who or what could that be? It must be the mind! But then we can no longer escape the thought of "a human being in the machine." We may have once thought that human thinking would be incomprehensible without the computer, but now the reverse appears—namely that the computer cannot think if there were no human being behind it. The story that started with "the machine in Man" turns out to end with "Man in the machine." I doubt that the proponents of "a machine in Man" will accept that turn around.

It is very easy in this discussion to fall into a terminological trap. One of the problems is that the word "information" has a specific meaning. Saying that both computers and brains process information is quite confusing. It is not the brain that creates information—the mind does. To say that my brain processes information when I read a book, for example, ignores the fact that a book only contains information for those who interpret and understand the code. The same can be said about the brain: it can only hold information because the mind is able to understand the binary code of the brain. In other words, humans are needed to understand the code. So, information only exists in the head of a user, or in the eyes of the beholder.

For now, we may conclude that a computer might be a good model of the brain, but not of the mind. That is, a computer may simulate what happens in the brain. After all, everything can be compared to everything else in some way. So, why couldn't brains be like computers? In other words, a computer is a model of the brain—nothing more or nothing less.

Yet, we keep facing a mystery. The enigmatic relationship between the mind and the brain is far from solved—and may never be for mortal beings. This relationship goes back to a much more important relationship, the one between body and soul, for the mind is the intellectual part of the soul. Perhaps St. Augustine can help us to understand the relationship between body and soul a bit better when he spoke about it in one of his sermons, centuries ago:

> *They questioned the very thing they themselves carried around with them; they could see their bodies, they couldn't see their souls. But they could only see the body from the soul. I mean, they saw with their eyes, but inside there was someone looking out through these windows. Finally, when the occupant departs, the house lies still; when the controller departs, what was being controlled falls down; and because it falls down, it's called a cadaver, a corpse. Aren't the eyes complete in it? Even if they're open, they see nothing. There are ears there, but the hearer has moved on; the instrument of the tongue remains, but the musician who used to play it has withdrawn.*

# Chapter 13

## A Matter of Fact

We are constantly being bombarded with *facts*, and this book is no exception. We want facts, not fiction. We want facts, not opinions. We want facts, not myths. We want facts, not emotions. We want facts, not lies. The most reliable facts are supposed to come from science. You must "follow the science," scientists say, so that you know the facts. We want scientific facts, not science fiction.

Scientists like to confront "dissidents" with the facts. Solicited or unsolicited, they tell us exactly how the world works and how the human body works. Every day we are being told that a new scientific fact has been discovered, for example, that the ozone layer is thinning in certain places, or that there is a gene for color vision on the X chromosome.

But what about those facts? What are the facts behind the facts? Scientists usually don't tell us that story. That's not their job anyway. Facts are facts to them, and that's it. Yet, we should study them under the microscope of philosophy.

### *The hard facts?*

It turns out that facts are "in fact" strange "things," once you think about them more deeply. On the one hand, it's true that facts have always existed, with or without the presence of human beings.

The fact that Planet Earth is not flat has always been a fact. Facts are facts, always and everywhere. In that sense, there have always been facts, for facts never change. Facts are supposed to be detached from time and space; they are true regardless of who you are and when and where you live; they appear as objective, absolute, and universal entities. In that specific sense, facts have always been there, no matter whether humans were around or not. Facts are facts—whether we like them or not, whether we know them or not, whether we discovered them or not. Facts are facts forever, so to speak. In that sense, they may be called "hard facts."

On the other hand, facts do not exist the way material things around us exist, because facts are not material things—they are mental entities. You cannot bump into them like it is with stones. In that sense, facts are not "hard facts." Whereas things and events are "physical" parts of our world, perhaps even rock-solid, facts are "mental" creations that cannot be physical or rock-solid (which does not mean, by the way, that facts only exist in the mind). A "fact" is not like a "thing" you can kick at. We are indeed surrounded by concrete things, events, situations, and processes, but facts do not belong to this list—they are of a different nature.

True, animals also live in a world of things, situations, events, and processes, but they don't live in a world of facts. Why not? Facts cannot exist without *concepts*. Actually, a "fact" is another concept of its own, and concepts are a unique part of human beings, as we discovered earlier. In other words, facts are uniquely tied to human beings. As soon as we start thinking and talking about them, facts seem to enter into space and time, because think-

# Chapter 13: A Matter of Fact

ing requires thoughts (rationality), and talking requires statements (language). And that's what makes the situation so complicated, because now we end up with at least four rather disparate elements: events, thoughts, statements, and facts. How are they related then? Let's find out.

To begin with, facts are not the same as events or things, the latter of which most people regard as the objective "hard-core" elements of this universe. Things and events may seem the best candidates to offer us a rock-solid foundation for our facts, but can they really fulfill their promise? True, things either do exist or do not exist, and events do happen or do not happen; you can ignore things and events, but you cannot deny them. So, by replacing facts with events, we might think we have found the solid objective foundation that we would like to have for our facts.

However, facts are very different from *events*. Unlike facts, events are dated, tied to space and time, whereas facts are detached from space and time. It is even considered a fact that certain events did not occur at all—for instance, it is a fact that Darwin did not have a copy of Mendel's 1866 article in his collection of papers and books (there is no event to point at here). Apparently, a fact is not the same as an event; the best we can say is that a fact is a *description* of an event, but not the event itself—which makes quite a difference.

But if it is true that facts are different from events, things, situations, and processes, then this might suggest that facts must be merely *thoughts*, existing only in our minds as something purely subjective. However, thoughts cannot be equated to facts either.

Thoughts can have some peculiar characteristics such as being imaginary, illogical, confused, time-consuming, and so on—whereas facts cannot. Facts, on the other hand, deal with what the events actually are, and not with what they might be. Facts are true, even if some people have never thought of them. Facts are always about something outside our thoughts and refer to something independent of our thinking. This makes facts unchangeable entities. True, sometimes we may declare something a fact, which turns out on further investigation not to be a fact, but the facts themselves cannot change. Therefore, a fact is not just a thought, but it may be the *object* of a thought.

What we just said—that facts are true, even if some people have never thought about them—may seem to contradict what we said earlier—namely, that facts only entered the world when humans came along. But that is only seemingly so. Facts are true, always and anywhere, but they could not be expressed until humans arrived on the scene. The reason for this is that facts are non-material entities which require a human mind. Once we start thinking and talking about them, facts seem to come back into space and time. That's where we need to address a fourth element in this discussion: statements. (Of course, one could assume at any time that facts have always existed in God's mind, but that is not the point of discussion here.)

Not surprisingly, some have claimed that facts are identical to what people say about them—that is, identical to (true) *statements*. However, if that were the case, there would be as many facts as there are statements (for instance, facts would be different in Eng-

## Chapter 13: A Matter of Fact

lish and Dutch). The fact that the earth revolves around the sun can be expressed in many ways—that is, in many languages, with simple or difficult words, with short or long sentences—but they do not change the facts.

Obviously, facts must be clearly distinguished from statements. Statements can be hypothetical, inaccurate, exaggerated, long-winding, difficult to understand, and so forth. Facts, on the other hand, cannot be any of these; a fact may be hard to accept, but never hard to understand; it is never hypothetical or half-true. There are even facts which everyone has forgotten, or which were never thought of yet, or which were never expressed yet in a statement. Therefore, we must come to the conclusion that a fact is not a statement, but it may be the *content* of a statement.

From this quite boring analysis follows that we are facing here a rather intricate situation: If facts are neither events nor thoughts nor statements, what then are they? A fact turns out to be a more complex concept than it may appear on the face of it. It certainly is not a "solid" entity like rocks and stones. Facts actually feature as a focus point at the intersection of those three other elements: A fact is not an event but the description of an event, not a thought but the object of a thought, and not a statement but the content of a statement.

That's quite a mouthful. But such a complicated analysis from under the philosophical microscope makes it immediately clear that facts are not just there for the taking. They are also part of our thinking and speaking. They have something that comes from us as well as something that comes from the world around us. How can that be? Well, facts are closely connected to these three other elements through the process of *interpretation*: Facts are interpretations of events by means of thoughts and statements. It is through interpretation that thoughts and statements transform events into facts. Facts need *events* so they can be tested; they need *thoughts* so they can be understood; and they need *statements* so they can be communicated. I hope you are still with me, for there is a lot at stake here.

### The naked facts?

What all the above makes compellingly clear is that there are no "naked facts." Facts do not exist without interpretation. It should not surprise us then that a camera, for example, cannot capture facts—all it can capture is things, situations, and events. To "capture" facts, we need at least concepts.

Thanks to concepts, we can interpret what we perceive. Let's take an example. When we describe what we see in the sky as "Those are moving dots," we use very vague concepts—namely, "dot" and "move"—and therefore very little interpretation. But when we say, instead, "Those are flying birds," we use more explicit concepts—namely, "bird" and "fly"—and therefore implement a more specific and detailed interpretation. And when we say, "Those are migrating geese," we inflate our interpretation even further with more explicit concepts—namely, "geese" and "migrate." Obviously, animals do not—and arguably cannot—some up with this. If they spot a hawk in the sky, they are not identifying the bird with a concept ("hawk") but merely as a signal of imminent danger that may require direct action. They don't study birds in the sky like bird lovers or ornithologists do.

In other words, again, so-called naked facts do not exist. We can try to keep interpretation to a minimum, but facts never allow themselves to be stripped "to the bone." We can reduce the amount of interpretation about "geese" in the sky, for instance, by replacing "geese" with "birds," or even further down with "dots," but this weakened interpretation inevitably also leads to a loss of information. We can strip down such a fact even further by claiming that spots appeared on our retinas. But that is a very thin interpretation of what "in fact" happened. Those who take a "safe" position have little to answer for, but also have little to share. The degree of certainty and the degree of information are inversely related to each other. Facts with a minimum of information are basically worthless. Facts like that never make it to the newspapers, let alone to scientific journals.

Scientists like to stick to the "naked" facts, and yet they want to put a lot of scientific information into them. Nowadays we would say that scientific statements are heavy with theory—they are theory-laden. Even a simple statement such as "this is a bloodstain" carries a good dose of theory. The built-in interpretation goes far beyond stating that red spots are blood spots, because on stage most red spots are not from blood, but from ketchup. To test whether they are really blood stains, we will have to rely on a good amount of theory—for example, by demonstrating the presence of iron in the stains, because it is a scientific fact that mammalian blood contains the element iron. We can even go so far as to determine the presence of the blood protein hemoglobin by applying an antibody-based test. Thus, the fact that a particular spot is a blood spot appears to be closely related to scientific concepts such as iron, mammal, hemoglobin, blood cell, and antibody.

In short, a certain amount of theoretical interpretation is packed into the observation and in the way we present the facts. Naked facts do not exist—although we will sometimes deem certain facts as so basic that we accept them as unquestionable without any forther doubt. They form, as it were, the piles in a marshy soil, on which scientists can then construct a solid structure of new theories and facts. Take the fact that humans are a product of their DNA. This fact can only be understood if we are familiar with the concept of DNA and with all the theories that have been built around it. The concept of DNA does not exist without those theories; that is, without those building blocks, the said fact would collapse like a house of cards.

### *Mere observation?*

Facts do not stare us in the eye. Because they require interpretation, they must be "captured" by the human mind. Just keeping your eyes open does not reveal any facts. We need more: interpretations! The way we perceive the world and all that is in it is, as we said earlier, described as "theory-laden." The "theory" part of this expression is probably overstated; it merely stands for the assumptions and expectations that steer our perception and observation. A

## Chapter 13: A Matter of Fact

statement about the world is said to be theory-laden if it presupposes or rests upon an assumption (a "theory")—in other words it is not a pure observation. The observation that someone has hypertension, for example, is "theory-laden" in the sense that this assumes several assumptions about blood pressure and about how we measure it.

There are at least two sides to being "theory-laden." One side is that perception is not a passive but active process. It is not a matter of just keeping your eyes open, but it also requires interpretation. Interpretation of what we see, perceive, and observe makes a big difference. For example, you may have seen many X-ray images, but by just looking at them you may not be able to interpret them properly. Or take this example: you may have seen many times that things on earth fall to the ground. Each time you see this, you see the action of gravity. Seeing gravity taking place is not really perception in the strict sense. In order to perceive gravity, one needs the concept of gravity, which enables us to see what no one had been able to see before Isaac Newton. The concept of gravity allows us to "see," for example, the similarity between the motion of the moon and the fall of an apple.

The other side is that perception may not even be perception in the literal sense. Let me just use a few examples from the history of science to make my point. When William Harvey proposed his theory of a closed blood circulation, he postulated capillaries. However, he could not *see* capillaries, because there were no microscopes yet. Or when Galileo Galilei proposed heliocentrism, he needed to see stellar parallaxes—a shift in position of a star observed from the earth on one side of the Sun, and then six months later from the other side. However, he could not *see* stellar parallaxes, because his instruments were still too crude. Or when Ignaz Semmelweis tried to stop the spread of childbed fever in his hospital, he spoke of "cadaveric matter" carried by students' hands from the corpses they had studied to the women they were treating. However, he could not *see* the germs of infection, because microbes had not been made visible yet. Or when Louis Pasteur proposed his germ theory, he did not *see* germs until Robert Koch made them visible under the microscope. Or when Urbain Le Verrier studied the strange motion of Planet Uranus, he postulated some large, unseen object that must have affected the planet's orbit. Calculations showed him that it would have to be a planet as large as Uranus and even farther away from the Sun. However, Le Verrier did not *see* planet Neptune until some of his colleagues in Germany had the courage to direct their telescopes to the spot where Le Verrier had told them to look.

Examples like these show that objects became only visible after they had been captured by interpretation. In other words, there is no such thing as "seeing in a neutral way" or "observing without expectation." This makes mere observation not a very productive starting point for scientific research. Even the idea of similar observations makes only sense if we know already what the assumed similarity is based on. Even if we assume that observations are the same for everyone—which they are not, except for images on the retina—we still need to acknowledge that observation statements involve theories of various degrees of sophistication, which makes them theory-laden to some degree.

## Can facts be separated from values?

Many today will say there are facts, and then there are values. The main difference between facts and values seems to be that facts are neutral and objective, whereas values are personal and subjective. That may seem common sense these days, but it raises many questions.

This position drives a wedge between facts and values, making for an unbridgeable gap between them. The general understanding seems to be that (scientific) facts are *descriptive* (true or false), whereas (moral) values are *prescriptive* (good or bad). This implies also that facts are seen as objective, whereas values are seen as subjective. As a consequence, facts are declared to be *value-free,* and values are declared to be *fact-free.* Before we continue, we must stress that this division is a relatively recent development. So, we should find out first how facts became value-free and how values became fact-free.

1. How did facts become value-free? Scientists have always argued that facts should be about reality, and not about what we wish reality to be. Whether a researcher is a Marxist or capitalist, socialist or liberal, sexist or feminist, believer or atheist, he or she will have to arrive at the same facts. What kind of research researchers choose may vary and be colored by their views, but the outcome of their research, the facts, would be the same for all of them. For ex-

ample, the fact that the earth revolves around the Sun is accepted by all astronomers nowadays regardless of their personal background. There is even general agreement on more controversial facts such as the claim that all humans are a product of their DNA. Whether that is all there is to the role of DNA is another question, but the genetic role of DNA is now accepted by all serious researchers as a fact. In short, scientific facts are shared by people of different beliefs.

This presumably makes for a sound distinction between fact and fiction. The choice of research may be personal but that should not be the case for the *outcome* of research—the scientific facts, that is. It is generally understood that science is looking for the truth—not the truth as we personally would like it to be, but the truth as it is. Scientific research must be guided by neutrality and objectivity, more in particular value-neutrality. Perhaps our own wishes or the wishes of others determine what we are going to search for, but not what we are going to find. Scientific facts can only be called scientific thanks to neutrality and autonomy. In all these cases it is about the "hard" facts, regardless of the wishes and interests of individuals or groups. Therefore, scientific researchers are supposed to be unbiased, autonomous, neutral, objective, and value-free observers. The facts they find in their research must be free of "value judgments." In short, the facts they find must be value-free as a result.

The term "value-free research" was introduced quite recently by the sociologist Max Weber (1864-1920). He applied this principle to research in sociology, but it has been widely applied to all scientific research since. That in itself makes for a healthy and sound view of scientific research. It creates a "division of labor" between science and other human enterprises. For example, the natural sciences investigate natural phenomena, not moral or religious issues. There is nothing pleading against that separation.

It only started to get confusing when people like Immanuel Kant, but also Max Weber himself, started to get involved in the division. They applied major border corrections, almost unnoticed. According to them, the boundary was no longer between nature and non-nature, but between the facts of what is and the values of what we want the facts to be. The objective facts would constitute the monopoly of the natural sciences, and the rest was considered subjective—thus, in a scientific view, a no man's land. The separation of fact from fiction had now become a separation of fact from *emotion*. From then on, values were seen as mere products of emotions, whereas facts had to be protected from such emotions.

2. How did values become fact-free?

This was another process which probably started with the philosopher David Hume (1711-1776). He made a distinction between factual judgments and value judgments. Whereas factual judgements make true statements about objects of the world, value judgements supposedly do not make true statements about the world but merely serve as expressions of subjective preferences and desires.

Based on this distinction, Hume declared that factual statements cannot logically entail value judgments. He famously stated that we cannot derive how things ought to be (a value) from the way things are (a fact). Put more technically, *ought*, a value, is not logically entailed by *is*, a fact.

Some have tried to get around this problem by defining moral terms in purely natural terms. But whoever tries to do so commits, with a term introduced by British philosopher G. E. Moore (1873-1958), a "naturalistic fallacy," which is the erroneous idea that what is natural can be defined as good in moral terms. Seeking pleasure for pleasure's sake, for example, may be natural but is not necessarily something that is also morally good. Sometimes this idea is generalized by declaring that anything found in nature is good, and then uses statements such as "Something is natural; therefore, it is morally good" or "This property is unnatural; therefore, this property is undesirable."

However, we should always ask the question as to whether things like "being natural," or "being pleasant," or "being desirable" are the same as "being morally good." Of course, we can equate them by mere definition, but that brings us back to what we wanted to equate. A moral concept cannot be redefined in non-moral terms, for then it loses its specifically moral aspect. By "redefining" morality in non-moral terms, we inevitably lose its distinctive moral character. Besides, all such attempts still move from an "is" to an "ought" statement, which is considered a fallacy since Hume. There is supposedly no way to connect moral values with natural facts. Thus, we end up with fact-free values.

3. Are facts really value-free?

It is quite paradoxical that the request for value-free facts is a value in itself—the value of objectivity. And this is not the only value scientists are supposed to respect. They must also strive for academic freedom, neutrality, and autonomy from external demands, all of which are again *values*. Moreover, scientists often have the tendency to demand—in addition to objectivity, autonomy, and neutrality— a monopoly position for their scientific claims. As if there were nothing in this world but scientific rationality and objectivity! These are "in fact" values scientists are willing to fight for.

We are definitely dealing with values here, almost hidden behind highly respected ideals. But reality is often quite different. Nowadays, we realize all too well that science has almost become a slave to the large capital, which is only available for certain "desirable" projects. This is another "value," but now of a financial nature. There is a lot of money pouring into science these days. Scientists may be neutral and autonomous in finding scientific facts, but for their search they are dependent on others who are needed for financial support. Some projects are supported, others are not. Until recently, research for military purposes offered good opportunities. Now different sponsors are in high demand.

Besides, because we are dealing here with values, they may conflict with other values. For example, during Joseph Mengele's scientific research in Auschwitz, the value of academic freedom came into conflict with the value of human dignity. A similar ethical problem arises with the question whether human fetuses (derived from in vitro fertilization) may be used for scientific research. Moral questions abound here. The Catechism (2294) summarizes this well: "It is an illusion to claim moral neutrality in scientific research and its applications." Science cannot examine morality—it is beyond its reach—but morality should always be allowed to interrogate science. Just as there is good and bad science in a methodological sense, there is also good and bad science in a moral sense.

Therefore, we must conclude that the gap between facts and values is "in fact" a fabrication. Facts are human constructs, as we found out earlier. Now we know that this involves not only interpretation but also *evaluation*. Facts tell us something about the world around us, of course, but they also tell us something about ourselves—and much more than most people would like to admit. Just as there are no hard or naked facts, there are also no value-free facts. So-called bare and value-free facts are merely abstractions that have carefully erased their traces of interpretation and evaluation. With facts you should not only ask yourself *how* we know them, but also *why* we know them. It sounds like "Show me your facts and I'll tell you who you are." Certain facts are considered much more significant than others. Lack of oxygen, for instance, is a much more important fact for humans than lack of prestige. That is also the reason why most people consider "applied" or technological science more important than "pure" or academic science.

Let's put this discussion in a wider context by using a revealing analogy. If someone gives me a 100-dollar note, then I may call that a fact, but this fact requires a very specific context. Anyone who does not know the financial context of this transaction will not see a banknote, but rather a piece of paper merely changing hands. Just as banknotes only exist by virtue of a banking economy, so do scientific facts only exist because of the institution of science.

Indeed, scientific facts do presuppose a certain context. First of all, there is a strong conviction behind them that declares scientific knowledge as not a matter of feeling, but of rationality—it is a highly respected ideal. How can such an attitude to science be justified? It would be circular reasoning to defend rationality on rational grounds. So, the choice for rationality can only be made on irrational grounds; it is a matter of "faith" and "dedication."

To put this in more general terms, our facts are not only describing, but also prescribing. They want to dictate to us the best way to describe facts. And that prescribing seems to work well. It already starts at school. Our Western society has been "scientific" from the cradle to the grave, albeit with a few pockets of resistance here and there. But a person does not live on science alone. So, what is a person to do with all those scientific facts that are supposedly stripped of their values? Value-free facts, if they exist at all, are basically worthless. That is one reason why many fundamentalists (whether Jewish, Christian, or Muslim) have great difficulty with the biological facts of evolutionary biology. Values and facts have become so closely intertwined that they can no longer be unraveled. In short, value-free facts do not exist.

4. Are values really fact-free?

The idea that values are fact-free is usually defended by claiming that there is no way to derive values from facts unless we commit the naturalistic fallacy. Therefore, values have got to be subjective, merely based on personal preferences. But are they only subjective, and not objective at all? Is there really no way to go from "is" to "ought"?

Some philosophers have asserted that there is such a way. The American philosopher John Searle, for example, argues that from the statement "Jones promised to pay Smith five dollars," it logically follows that "Jones *ought* to pay Smith five dollars." The reason is that the act of promising, a fact, by definition places the promiser under obligation, a value. The Scottish American philosopher Alasdair MacIntyre argues something similar by asserting that from the factual statement "This watch is grossly inaccurate" validly follows an evaluative conclusion, "This is a bad watch."

All this may sound rather peculiar and farfetched. However, it has important implications. Think of St. Thomas Aquinas' view that the *natural law* of what God commands us to do can be known by everyone through our human nature—which means everyone can know what humans *ought* to do from the way they *are*. Nowadays, not surprisingly, this way of thinking has come under attack as violating the rule that we cannot derive what ought to be done from the way things are—that is, "ought" does not flow from "is." The philosopher Jeremy Bentham, for instance, criticized the Aquinian natural law theory because in his view it was based on a naturalistic fallacy, claiming that it derived how things ought to be in morality from the way things are in nature.

In response to this attack on Aquinas, the late Notre Dame philosopher Ralph McInerny explains that "ought" is already bound up in "is," in so far as the very natures of things have ends or goals within them. A clock, for example, is a device used to keep time, so because it "is" a clock, the clock "ought" to keep the time. In like manner, if one cannot determine good human action from bad, then one does not really know what a human person is by nature. In a similar vein, Pope Benedict XVI reiterated recently the Aquinian view that "the ought does flow from the is." What he meant is that once we get a sense of who God is and what a human being is—created in God's image and likeness—certain "oughts" do flow from that.

Therefore, facts are intricately connected with the world of "oughts," which is basically the world of morality. It is essential to all of us that we can discern if human beings are doing the right thing or not in moral terms. Whereas human movements are subject to physical constraints, human actions are subject to moral ones. So, the conclusion of this chapter must be that there are no hard facts, no naked facts, but also no value-free facts. Facts are more intricate and complicated than they appear in the eyes of many.

# Chapter 14

## The God of Facts and Values

In the previous chapter, we demonstrated that facts are very different from events, thoughts, and statements. It is through interpretation and evaluation that thoughts and statements transform events into facts. Facts are what we express in statements with the help of concepts.

Thus, the question arises as to where these concepts and all statements about facts come from. The problem is that they can't come from the animal world, nor from mere observations, nor from our languages, nor from our surroundings, nor from our brains with their neurons, nor from our genome with its genes. So, where else, then, could they come from? Where do they reside? They must exist somewhere for them to be available to each one of us individually and to all of us together. But where could that "somewhere" be? Amazingly, many scientists don't seem to be aware, at least not consciously, of the enigma that we have here.

### *What concepts are not*

Some think that concepts are just labels for what we can find defined in dictionaries. However, ultimately, dictionaries must always use circular reference since all words in a dictionary are defined in terms of other words. Dictionaries can never step outside their own confines to refer to something outside of the dictionary. No matter how hard we try, we will never be able to get the concept that we try to describe and define "off the ground"—every trial falls back on other concepts. Even if we decide to stop this endless regression by declaring a few concepts as the pillars that carry the rest of the conceptual framework, then we still need to explain where those so-called fundamental concepts come from.

Perhaps concepts come from observations then? We can explain the concept of snow by pointing at the snow we see. Right? However, that cannot be a solution to our problem. First of all, for certain concepts there may be nothing to point at. To explain the concept of "tomorrow," for instance, there is nothing to point at. Second, pointing at something like a cell under the microscope does not automatically generate the concept "cell." The history of cell biology shows us how tedious its discovery was. It's only when we know the concept of a cell that we can point at a cell and identify it as a cell. Anything can be pointed at, once it has been identified; but not everything that has been pointed at can be identified with a concept. Third, showing observers several white things, for instance, does not make them see or identify whiteness; white things have many other things in common. In other words, pointing at a set of white things does not automatically generate the concept of whiteness. To determine what qualifies to be in the set of white things presupposes the very concept, whiteness, which it is meant to explain. In other words, observation does not create concepts but utilizes them.

## Chapter 14: The God of Facts and Values

If abstract objects such as facts, values, and concepts—plus all the statements that express them—do not depend for their existence on the material world nor on the human mind, then there is only one rational option left: they must exist in a "third realm" that is neither material nor mental. This idea is usually associated with the Greek philosopher Plato. However, Plato's position faces multiple, rather technical problems, which we will not discuss here. But there is a much more acceptable version of this third realm—arguably the only valid one—which goes basically back to St. Augustine, and was later elaborated by the philosopher Gottfried Leibniz, and then strongly backed by the logician Gottlob Frege, who wrote in 1918 that "a third realm must be recognized" —a realm of meaning or sense, independent of anyone's individual thoughts. In this version, abstract entities or objects do indeed exist, but they can do so only in an infinite, eternal Divine Intellect. As St. Augustine put it, there must be something "that all reasoning beings, each one using his own reason or mind, see in common."

This probably raises the question as to why abstract entities can't exist in *human* intellects, instead of one *Divine* Intellect. The reason is that human intellects are contingent—they do not have to exist but come into being and pass away. If abstract entities existed only in human intellects, they would have to come into existence and could go out of existence again. Besides, how could we have them in common with other human intellects? And more importantly, before humanity emerged, there would and could not have been any abstract entities—which implies, for instance, that snow wasn't white then. Of course, snow has always been white, but its whiteness became visible to human beings through the concept of "whiteness." These abstract entities—concepts, statements, facts, and values—are *objective* realities that exist outside our minds in a world of their own. Call it a "third world," if you like.

## The Divine Intellect

Consequently, the only sort of intellect on which abstract, universal, and timeless concepts, statements, facts, and values could ultimately depend for their existence would be an intellect that could not possibly have *not* existed—a Divine Intellect, the Mind of God, the First Cause. It is this Divine Intellect that grasps and holds all the logical relationships between all facts and between all statements with all their universal concepts. This is an Intellect that eternally understands all *actual* truths and all *possible* truths, as well as all *necessary* truths. It is God who causes the world to be such that a statement is either true or false, either possible or impossible, either necessary or unnecessary. This is so because facts, values, statements, and their concepts exist as thoughts in the Divine Intellect.

To put it in a more charged way: without faith in God's Divine Intellect, we have nothing to claim as truth. We are entitled to say that the statement "snow is white" is true only if snow is indeed white in the Mind of God. One of the ways to find out whether something is indeed in the Intellect of God is by "reading God's Mind"—either in nature or through reason. We do so, for instance, when we "interrogate" the universe through investigation, exploration, and experiment, as well as when we use logic, reason, and philosophy.

Only the Divine Intellect can make it possible that you and I share the same fact when we say and think that snow is white or that mutations are random—even if the statement of these facts differs in English, German, or whatever language. Only the Divine Intellect can explain that the world is an *objective* and orderly entity knowable to the human intellect, which is also an orderly and objective product of the rational and consistent Divine Intellect. The book of Genesis could not have put it more simply: "God created man in his own image, in the image of God he created him." God created us with the desire to know and learn and understand that which exists in His Intellect.

As we showed earlier, facts and values are neither events nor thoughts nor statements, therefore they can only exist in the Mind and Intellect of God. That's where their truth must originate. The late astrophysicist Sir James Jeans once said, "The Universe begins to look more like a great thought than a great machine." The "great thought" that Jeans speaks of here is not just a thought of someone's human intellect but must be a "thought" of the Divine Intellect. The Divine Intellect is where concepts reside, but it's also where facts and values come from. Indeed, God has given us the tools to "read" His Mind and to begin to think a bit more like Him.

## Objective knowledge

I consider it quite striking that—not related to the previous reflections—one of the twentieth century's greatest philosophers of science, Karl Popper, expressed a similar view when he introduced his concepts of "objective knowledge" and "knowledge without a knower." He once wrote, "Knowledge in the objective sense is *knowledge without a knower*; it is *knowledge without a knowing subject*" (italics are his). He is in fact claiming that objective knowledge does exist, even without a human knower or a human knowing subject. That "snow is white" was true even before there were any human beings or "knowing subjects" around. But there is one caveat: can *knowledge without a knower* still be knowledge? That would easily become a paradox.

There must be a knower for it to be knowledge, but perhaps that knower is not a human knower. Therefore, we must conclude that there *must* be a knowing subject somewhere *outside* any individual human minds and *prior* to them. This "knowing subject" *must* be the Divine Intellect. In other words, God is the "knower," the "knowing subject," behind all objective knowledge, including the knowledge in individual human minds. This means there cannot be any knowledge without the knowing subject of God. Only this can explain why scientists can do their work of exploring and investigating nature.

What all scientists are "in fact" doing through their research is reading the Mind of God—often without them knowing it, let alone acknowledging it. Stephen Hawking once put it this way:

finding a unified theory in physics would be "the ultimate triumph of human reason—for then we would know the mind of God." Indeed, God has given us the tools to "read" His Mind and to begin to think a bit more like Him. In some mysterious way, our human intellect is able to capture concepts, facts, and values that reside in God's Divine Intellect. Only the Divine Intellect can explain that the world is an objective and orderly entity knowable to the human intellect, which is also an orderly and objective product of the rational and consistent Divine Intellect. Thanks to this connection between God and the universe He created, between God's Intellect and what our human intellect knows about the world through concepts, statements, facts, and values, the universe becomes more intelligible and sensible.

# Index

## A

academic freedom ............. 167
Aquinas, Thomas .. 77, 81, 170
Aristotle .......... 34, 37, 44-6, 54
Artificial Intelligence .... 145-7
Augustine ...... 21, 85, 151, 175

## B

Barr, Stephen ..................... 141
Benedict XVI 57, 87, 116, 171
binary code 137-8, 144-5, 149
Bohr, Niels ........................ 3-4
Bonhoeffer, Dietrich ........... 17
brain ........... 1, 13-5, 75, 78, 93, 100-1, 112, 135-6

## C

causality ....... 3-4, 10, 13, 16-7, 19-21
chance .......... 20, 58-61, 66, 98
Chesterton, G.K. ................. 82

coincidence ..... 58-62, 64, 66-7
computer ..... 15, 69, 89, 136-7, 139, 141, 144-50
conception ................. 107, 109
concepts .......... 64, 126-7, 131, 142-3, 146-7, 154, 158, 173-9
creation ........ 17, 19-21, 70, 79
creation theory ............... 79-80
creationism ..... 79-80, 83, 85-6

## D

Darwin, Charles ... 7, 21, 51-2, 58, 68, 70, 73-5, 81, 83, 119, 155
Darwinism ........................... 83
De Duve, Christian ............. 45
Deism ............................ 17, 19
Descartes, René ............. 133-5
determinism ...... 3, 18-9, 92-5, 105

DNA..........v, 35-6, 41, 43, 54, 69-70, 73-4, 89, 92, 100, 111-2, 138, 160, 164

Dobzhansky, Theodosius...79

**E**

Einstein..............................2-4
evaluation...........................168
evolution......1, 6-7, 16, 21, 31, 51-2, 57-68, 77-86, 121
evolutionism........81-3, 85, 86

**F**

Fall.............................80-1, 87,
feedback...................23-5, 28-9
First Cause....................16, 176
Francis..............................116
free will....13-5, 18, 89, 92, 94, 105, 113
functionality.............10, 19-21

**G**

Gaia................................28-32
gay gene............96, 98-99, 101

gender....................107, 109-10
gender dystrophia.............114
genderism..........................118
genetics........88-90, 92, 96, 99, 105, 111
god gene......................99, 101

**H**

Hamer, Dean....96-102, 104-5
Hawking, Stephen.............178
holism............................34, 36
homosexuality.96-8, 112, 114
Hume, David..................165-6
Husserl, Edmund..............142

**I**

intellect..........77, 146-9, 175-7
intelligence.....................145-7
intentionality...........15, 140-1
interpretation............. 4, 9, 46,

**J**

Jeans, James......................177
John Paul II........................57

## K

Kant, Immanuel .... 5-6, 9, 165
Kircher, Athanasius ............ 46

## L

Laplace, Pierre Simon ......... 18
Leibniz, Gottfried ............... 175
Lewis, C.S. ............................ 86
Lovelock, James ............... 28-9

## M

machine ......... 133-6, 140, 145, 148-50, 177
MacIntyre, Alasdair .......... 170
macro-evolution ............ 62, 64
McHugh, Paul .................... 117
McInerny, Ralph ............... 170
meaning ................. 17, 67, 126
Medawar, Peter ............ 91, 95
mind. 13-5, 93, 140-2, 146-50, 154-6, 160, 175-9
model .... 9, 29-30, 72-5, 135-6, 150
Moore, G.E. ........................ 166

morality ...... 87, 89, 94, 166-71
mutation ..... 59-62, 65, 78, 177

## N

natural law ......................... 170
natural selection 1, 6-7, 21, 24, 31-2, 60-2, 65-6, 74, 78-9, 82-4, 101
naturalistic fallacy 166, 169-70

## O

observation . 3-4, 160-2, 173-4
Original Sin ...................... 86-7

## P

Paley, William ................. 20-1
Pascal, Blaise ....................... 87
Pasteur, Louis ... 37, 45, 49-51, 162
perception .. 100, 146-7, 160-1
Pius XII ............................... 57
Popper, Karl ...................... 178
Pouchet, Felix ................ 49-51
Premack, David ... 123, 129-30

## Q

quantum mechanics ..... 1, 3, 7

## R

randomness .................... 60, 66
realism ........................... 2-4, 7
Redi, Francesco ................... 47
reductionism .............. 33-4, 36
religiosity ........................ 11, 15
reproduction. 21, 64, 107, 115
Russell, Bertrand ................ 90
Ryle, Gilbert ........................ 13

## S

Searle, John .......... 139-40, 170
self-regulation ......... 23-4, 26-8
semantics ................ 126, 144-5
signals .... 36, 40, 121-2, 125-7, 142
speciation ........................... 61
Spencer, Herbert ................. 83
spontaneous generation 45-51
statements ........ 155-65, 173-9
syntax ..................... 138, 144-5

## T

theory-laden .......... 160-1, 163
transgenderism .................. 118
twin studies .................. 99, 102

## U

understanding ....... 140-1, 147

## V

value-free ...................... 163-71
values ................ 12, 23, 163-79

## W

Weber, Max .................... 164-5

## X

X-chromosome ................... 97

## Y

Y-chromosome .............. 107-9

www.ingramcontent.com/pod-product-compliance
Lightning Source LLC
Chambersburg PA
CBHW070538090426
42735CB00013B/3013